AF848405

GEOTECHNICAL RISK IN ROCK TUNNELS

BALKEMA - Proceedings and Monographs
in Engineering, Water and Earth Sciences

SELECTED PAPERS FROM A COURSE ON GEOTECHNICAL RISK IN ROCK TUNNELS/AVEIRO/PORTUGAL/16–17 APRIL 2004

Geotechnical Risk in Rock Tunnels

Editors

António Campos e Matos
Dep. of Civil Engineering, Faculdade de Engenharia da Universidade do Porto, Portugal

Luís Ribeiro e Sousa
Dep. of Civil Engineering, Faculdade de Engenharia da Universidade do Porto, Portugal

Johannes Kleberger
iC consulenten ZT GmbH, Salzburg, Austria

Paulo Lopes Pinto
Dep. of Civil Engineering, Faculdade de Ciências e Tecnologia da Universidade de Coimbra, Portugal

CRC Press
Taylor & Francis Group
Boca Raton London New York

CRC Press is an imprint of the
Taylor & Francis Group, an **informa** business
A TAYLOR & FRANCIS BOOK

Sociedade Portuguesa de Geotecnia
Is especially grateful to

BRISA – Engenharia e Gestão
FCT – Fundação para a Ciência e a Tecnologia
FLAD – Fundação Luso Americana para o Desenvolvimento
Metro do Porto
Metropolitano de Lisboa
RAVE – Rede Ferroviária de Alta Velocidade
Universidade de Aveiro

CRC Press
Taylor & Francis Group
6000 Broken Sound Parkway NW, Suite 300
Boca Raton, FL 33487-2742

© 2006 by Taylor & Francis Group, LLC
CRC Press is an imprint of Taylor & Francis Group, an Informa business

No claim to original U.S. Government works

ISBN-13: 978-0-415-40005-3 (hbk)

This book contains information obtained from authentic and highly regarded sources. Reasonable efforts have been made to publish reliable data and information, but the author and publisher cannot assume responsibility for the validity of all materials or the consequences of their use. The authors and publishers have attempted to trace the copyright holders of all material reproduced in this publication and apologize to copyright holders if permission to publish in this form has not been obtained. If any copyright material has not been acknowledged please write and let us know so we may rectify in any future reprint.

Except as permitted under U.S. Copyright Law, no part of this book may be reprinted, reproduced, transmitted, or utilized in any form by any electronic, mechanical, or other means, now known or hereafter invented, including photocopying, microfilming, and recording, or in any information storage or retrieval system, without written permission from the publishers.

For permission to photocopy or use material electronically from this work, please access www.copyright.com (http://www.copyright.com/) or contact the Copyright Clearance Center, Inc. (CCC), 222 Rosewood Drive, Danvers, MA 01923, 978-750-8400. CCC is a not-for-profit organization that provides licenses and registration for a variety of users. For organizations that have been granted a photocopy license by the CCC, a separate system of payment has been arranged.

Trademark Notice: Product or corporate names may be trademarks or registered trademarks, and are used only for identification and explanation without intent to infringe.

Publisher's Note

The publisher has gone to great lengths to ensure the quality of this reprint but points out that some imperfections in the original copies may be apparent.

Visit the Taylor & Francis Web site at
http://www.taylorandfrancis.com

and the CRC Press Web site at
http://www.crcpress.com

Geotechnical Risks in Rock Tunnels – Campos e Matos, Ribeiro e Sousa, Kleberger & Lopes Pinto (eds) © 2006 Taylor & Francis Group, London, ISBN 0 415 40005 8

Table of Contents

Geotechnical Risks in Rock Tunnels – Campos e Matos, Ribeiro e Sousa, Kleberger & Lopes Pinto (eds) © 2006 Taylor & Francis Group, London, ISBN 0 415 40005 8

Preface

The theme of Geotechnical Risk in Rock Tunnels is certainly challenging and absolutely actual. As the number, length and depth reached by the new tunnels being designed and constructed through out the world increases, so do the difficulties and risks involved. Furthermore, the financial conditions supporting the entrepreneurships are becoming more rigid, making the advanced assessment of cost and level of risk associated to the loss of control of this cost a critical issue.

Recognising the importance of this topic, a Course on Geotechnical Risk in Rock Tunnels was held at the University of Aveiro, Portugal, in April 16–17, 2004, under the support of the Portuguese Geotechnical Society (SPG). Renowned specialists, from nine different countries, presented contributions on several aspects of risk management and control that are now published in this book. The course was attended by close to two hundred people, from fifteen countries and the exchange of experiences was very rewarding. The editors would also like to express their recognition for the excellent organization and logistics put together by Prof. Claudino Cardoso, from the University of Aveiro.

A main target of the Course on Geotechnical Risk in Rock Tunnels was to shed light on the issue of risk in tunnelling, to give guidance concerning techniques and tools available, on which procedures to follow during the various phases of a tunnel project. Relevant risks in tunneling have to be identified, characterized and minimized by careful design. Every risk identified has to have an owner, usually the client or the civil works contractor. During construction a risk management process has to be employed in order to deal safely with the remaining risks.

The contribution by L. Ribeiro e Sousa presents what may go wrong in underground projects if risks are not considered carefully. Lessons learned from accidents and failures are discussed.

S. Babendererde, E. Hoek, P. Marinos and Silva Cardoso report on Geological Risk in the use of TBMs in heterogeneous rock masses. Their experience on the complex project of "Metro do Porto", where a difficult geology was overcome successfully using and Active Support System, is presented.

H. Einstein explains the use of Decision Aids for Tunnelling, and its successful application on a number of projects.

N. Barton gives the reader insight into his worldwide experiences in fault zones and TBM tunnelling.

J.M. Rodriguez Ortiz reports Geomechanical problems in recent Spanish tunnels. A review is made concerning problems of conventional and mechanised tunnels in hard rock, as karstic ground, roof collapse, wearing of cutting tools, evaluation of support needs, overexcavation, etc. More specifically are dealt with the problems related to soft rocks, especially the theoretical and practical design of tunnels in swelling and squeezing ground as well as some cases where closure of the section and heaving of the invert occurred in soft, non-swelling rock.

It is well known fact, that poor or inadequate description of ground conditions is one of the typical reasons for excessive cost and time overruns in tunnel construction. Claims based changed ground conditions alone may reach 50 or more percent of the initial cost estimates.

In order to avoid unknown ground conditions, sound engineering-geological and hydrogeological works are indispensable. The article by G. Harer reports on the benefits of a target oriented site investigation for Koralm Tunnel (approx. 32,8 km long), making it Austria's longest tunnel and the seventh longest tunnel in the world, explains details of the site investigation approach. Performing site investigation step by step creates required knowledge and controls risks in order to save time and budget significantly by providing sufficient geological data for every project phase.

The paper of G. Venturini et al. explains why the hydrogeological risk may represent one of the most severe risks in rock tunneling showing examples of unforeseen high water inflow of >50 l/sec leading to suspension of tunnel construction of several strategically important tunnels as well as considerable financial losses, generating important claims due to the underestimation of such problems in tender documents. The paper also explains how to minimize such risks.

The contribution by I. Pöschl and J. Kleberger explains, how the world wide used ground classification approaches can accommodate for geotechnical risks by combining quantitative and qualitative classification methods. The characterization process proposed permits the definition of "rock mass types" (RMT) that display typical deformation/failure patterns and/or may impose specific geotechnical hazards.

J. Daller's contribution on the approximately 13 km long Wienerwald Railway Tunnel Project consisting of two single track tunnels, each about 10.900 m long, a 2.236 m long double track section with a 409 m long enlarged cross section at the transition to the twin tubes, a ventilation cavern, a permanent ventilation shaft, three optional temporary ventilation shafts, three permanent emergency exits, 22 cross passages connecting the single track tubes at every 500 m distance, inclined mucking galleries equipped with conveyor belts and a temporary construction access, explains how risk assessment and risk control were understood as most important aspects from the very beginning of the design process for a 360 Mio Euro civil works tunneling contract near Vienna.

P. Schubert explains how risks that could not be completely eliminated during design have to be dealt with during construction. Risk management in tunnelling is a continuous process from alignment selection until completion of a project. The usual steps of risk management, such as risk identification, avoidance or reduction during design development, risk sharing with the contractor, and controlling the remaining risk during construction equally apply to the tunnelling construction phase. Basic elements of the risk management methodology and case histories are presented.

W. Unterberger reports on noise and vibration, important environmental issues to be addressed when designing high-speed railroad lines. While noise prognosis follows well-established design procedures, vibration prognosis and mitigation is a much younger field, which is still very much open to innovation. In Austria, a system of vibration prognosis has now been practiced and successfully applied for several years at a number of high-speed railway projects that relies on a combination of measurements, analytical approaches and numerical analysis. The prognosis system is currently being expanded to include probabilistic approaches.

António Campos e Matos, Luís Ribeiro e Sousa,
Johannes Kleberger and Paulo Lopes Pinto

Geotechnical Risks in Rock Tunnels – Campos e Matos, Ribeiro e Sousa, Kleberger & Lopes Pinto (eds) © 2006 Taylor & Francis Group, London, ISBN 0 415 40005 8

An introduction to geotechnical risk in rock tunnels

António Campos e Matos
Dep. of Civil Engineering, Faculdade de Engenharia da Universidade do Porto, Portugal

Paulo Lopes Pinto
Dep. of Civil Engineering, Faculdade de Ciências e Tecnologia da Universidade de Coimbra, Portugal

ABSTRACT: The uncertainties in many of the areas related to tunnels in rock mass are well apprehended by most of the entities involved in the different parts of the process. Owners, designers, contractors, insurers, all agree that geotechnical works in general and tunnels in particular have a high degree of risk that needs to be assessed and managed. In this course the justification for the importance of specific education in this area is expressed and key aspects related to geotechnical risk are presented.

1 MOTIVATION

The first question that should be addressed is "Why is there the need for a course in geotechnical risk?"

If we look around, we will not find only civil engineers, geologists and engineer-geologists, but also economists interested in topics about financing. We will also find insurers interested in risk coverage. What have brought us here were not the specifics about tunnel design and construction, but a broader topic: what are the risks involved with these subterranean works, in the different aspects of the work, such as "construction and safety", "financing and control", "exploration and maintenance", and others. As it will be shown during the course, all these aspects are indeed interconnected with one another and particularly with the ground conditions.

Someone said that the decision of building a tunnel is the last one to take, and should be taken only after all other options have been pursued. Even though this may sound extreme, the risk associated with this type of works is always present and should not be neglected. The cost of a tunnel is very substantial and the possible concentration of this cost on a sole constructive procedure of low flexibility, takes the decision to levels of high risk.

The concept of risk crosses through the societies with significance and value that differ greatly in space and time. In civil engineering, common sense would associate the word "risk" with "accidents", or at least to a situation "seriously out of control". The concept of risk is reaching other areas such as financing and also exploration. In this field, maintenance costs and compensation due to delays in railway exploration have reached values of extreme importance. It is well known how often these questions are related to geotechnical problems. These risks cannot be ignored in the actual procedures of large highway and railway projects (Fig. 1).

Why a course about these risks, in tunnels or more broadly, in underground works, in rock mass? Why not on other types of construction and/or soil conditions?

Tunnels intersect and strongly interfere with natural materials, involving them as a structural component of their own stability. This interaction is much more significant than in any other civil engineering work. As natural rock mass is a potential source of enormous parameter variability, it is easily understood that risk assessment and management at every step, are almost imperative for such projects, as opposed to others where parameter variability is smaller and, most of all, controllable.

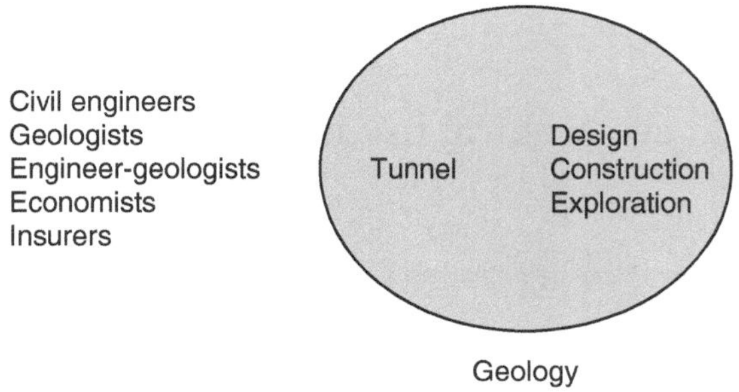

Figure 1. Parties involved on risk analyses in tunnels.

On other hand, soils are, due to their origin, usually more homogeneous than rock mass. This does not imply that tunnels through soil have necessarily less risks, it means that tunnels in rock deal with more variability in their characteristics. Some of these can be accounted by means of statistical procedures, but some are complete surprises and will not be treated by those procedures. Generically, rock mass present more anisotropy and heterogeneity.

Translating theses conditions, constructive and design procedures do not follow the same rules of other civil engineering works. That is a direct consequence of the need to control risks. Hence, knowing and accepting the fact there are no processes of identifying and characterizing the variables at play, with the same depth achieved for other constructions, there is the need to resort to other procedures of observation and verification during the construction, with possibility of corrections and adjustments of the project, which is an accepted way to control risk.

In underground works, and differently from other constructions, we deal with parameters that are highly variable. Their timely identification and characterization can never be complete. Therefore, the development of the works will always be associated to a variable number of uncertainties that can never be fully exhausted.

The understanding of these procedures by all the parties involved, and not only by the engineers, is absolutely necessary to the success of the works.

Throughout Europe, and also in Portugal, there are several large underground projects, in design or construction stages, mainly connected to new road and railway networks.

Portugal, in the near future, faced with a considerable number of new tunnels that by their dimension, number, and financial procedures, will require careful preparations for these challenges.

On the railways and in the case of high-speed networks, there are expected many kilometers of tunnels that will likely be built through rock masses. In Metro do Porto and Lisbon we may expect important tunnels to be built. Also underground activity will include tunnels built for water supply and new highway tunnels.

The funds available for these projects are extremely relevant, but they will not have the elasticity of times past. What is even more important and relevant than the global cost estimate for each work, is the effort to maintain the costs controlled during the development of the works.

Each underground work has its intrinsic cost, dictated initially by its characteristics and the market, and corrected during the construction phase by the geology encountered. It is not possible to influence this cost much. If the work is preceded by enough studies and risk is bound contractually, this cost will be reached with tranquility. In the opposite case, the same cost will be obtained, even surpassed, but with negative conflicts.

The key to cost control is the way studies are performed, in depth and extension, and the way the contracts consider these aspects. Nevertheless, the opinions about the best solutions are not necessarily uniform, changing from country to country, function of the legal framework of construction contracts (Fig. 2).

Transferring risk completely from the owner to the contractor seems an attractive solution, but it is by no means a flawless solution and quality and schedule can be strongly affected. Total risk

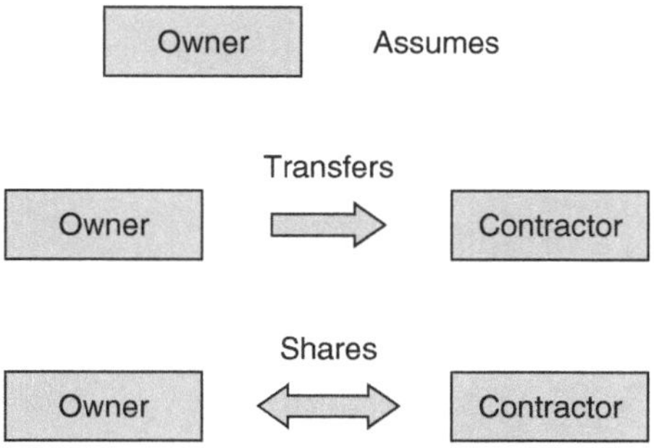

Figure 2. Options available when dealing with risk.

assumption by the owner is in practice difficult to accept because of public financing. Risk sharing is always an intermediate and appealing approach but not always efficient. Certainly, the less studies are promoted, higher will be the risks taken. We can see that it is not an easy task to be the decision maker in these projects. Certainly, the development of studies, the follow-up by consultants and the risk analysis are indispensable tools to frame risk within acceptable boundaries. The experience from other countries where this is a more common procedure will certainly benefit us.

2 RISK, AN ANALYTICAL AND EMPIRICAL CONCEPT DECISION IN GEOTECHNICAL ENGINEERING

The need to decide in an environment full of uncertainties and even surprises – typical of rock mass, leads to the adoption of design and construction methodologies very different from other disciplines.

In civil engineering, risk analysis balance between probability based and empiric approach. While the first may supply a numeric result of the evaluation – apparently ignoring experience, the latter bound risk by previous successes and failures – apparently ignoring the numerical analysis.

In fact, we all understand that on a phase of risk assessment, numerical results are the expected outcome. Nevertheless, facing the vast difficulties that have often to be dealt with in probabilistic analysis, the resort to experience is certainly welcome and required.

The empirical processes, when properly consolidated by the justification and understanding of the phenomena, are an indispensable tool for risk reduction. The fact that they do not allow numerical integration of risk does not curtail their interest and utility.

An empirical and non-numerical risk evaluation, based on experience will be, for instance, the risk comparison between the performances of two or more constructive solutions, in a known geological environment for which exist information and experience.

When following the path of probability based risk assessment, the inputs, for instance, are:

- Identification and statistical characterization of the variables (geotechnical parameters, strength, de-formability, joint orientation, etc), using Beta-type distributions or other;
- Use of techniques of correlated variable behavior and predefined statistical distributions (e.g. Point Estimate Methods – PEM);
- Choice of a Capacity–Demand Model (CDM)

The results of these analyses are curves of failure probability and the reliability index. Risk will be the product of these probabilities by the impact associated to the occurrence.

Risks can be decomposed in different aspects, being some of the most important:

- Constructive (accidents and solution of un-expected problems)
- Financial (lack of cost control)
- Exploration (direct and indirect)

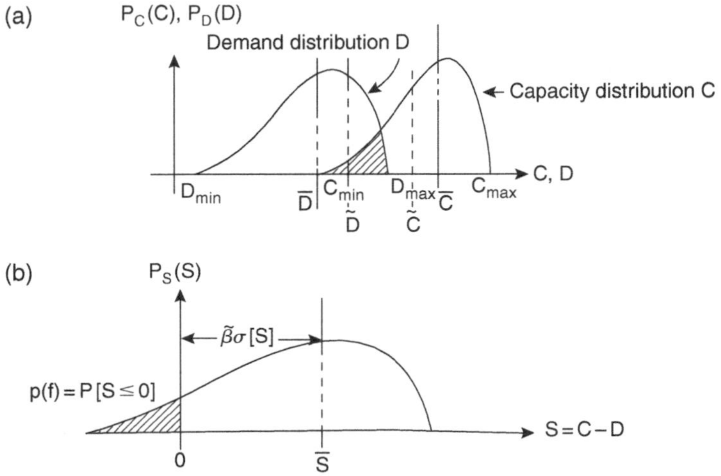

Figure 3. (a) Capacity–Demand Model; (b) Safety Margin and Reliability Index (McCraken, 1994).

The need to know the risks and decide based on this information, has introduced in soil mechanics, a long time ago, studies based on statistical knowledge. The introduction of reliability concepts and risk theory, together with the consequences associated to failures, gave origin to risk studies:

- Risk identification
- Risk characterization
- Risk quantification
- Risk assessment
- Risk management

3 THE GEOLOGY TIME

The crucial impact on underground works that geology carries, often unconsciously, should be discussed. After all, "Risk" results from the difficulty to deal with geological parameters.

At the time when geology was integrated on a descriptive basis activity, the responsibility and the associated risk were almost not identified. But when the geological parameters began being quantified, the range of possible errors or shortcomings and associated risks increased. This is basically positive, because it visualizes the problems.

The way geology conveys information to engineering has to be understood. A deterministic perspective of the problem is not enough. The variability associated to the parameters, the confidence and reliability of interpretation and characterization has to be an integrate part of the geological studies.

We consider it important to bring to Portugal the experience of those who deal with this demanding environment, in the midst of so many uncertainties. The balance, good sense, experience are all fundamental requirements, but also mechanics of material, statistics and reliability, detection technologies, characterization, and inspection.

Society expects from engineering an exemplar performance in these works, in order to achieve safe construction within the budget.

Geotechnical engineering expects that young engineers have the will and knowledge to understand that they are dealing always with variables of difficult identification and characterization, that models and computer codes are extremely useful means to understand the phenomena, but they should never be applied without been proved by the local realities and variability control.

We should never forget Karl Terzaghi's sentence: "*I am more and more amazed about the blind optimism with which the younger generation invades this field, without paying attention to the inevitable uncertainties in the data on which their theoretical reasoning is based and without making serious attempts to evaluate the resulting errors*".

We recommend strongly risk control but never the acceptance of the concept of a "risk free" geotechnical work.

REFERENCES

Cruden, D. & Fell, R. 1997. *Landslide Risk Assessment*. Rotterdam: Balkema.

Harr, M.E. 1981. *Méchanique des Milieux Formés de Particules*. Presses Polytechniques Romands.

McCraken, A. 1994. "*Reliability* based design using point estimate methods and capacity-demand model". In B.O. Skipp (ed.), *Risk and Reliability in Ground Engineering*, Thomas Telford, ICE, London.

Réthàti, L. 1988. *Probabilistic Solutions in Geotechnics,* Elsevier, Amsterdam.

Geotechnical Risks in Rock Tunnels – Campos e Matos, Ribeiro e Sousa, Kleberger & Lopes Pinto (eds) © 2006 Taylor & Francis Group, London, ISBN 0 415 40005 8

Learning with accidents and damage associated to underground works

Luís Ribeiro e Sousa
LNEC and University of Porto, Portugal

ABSTRACT: Most accidents and other related events are often associated with uncertainties. It is therefore essential to develop risk analysis systems and to avoid its occurrence. The occurrence of accidents and incidents in underground structures is not as unusual as in other geotechnical structures. However the dissemination of these events are not common due to the legal problems involved. There is a general the tendency to minimize the dissemination of the involved damages and their causes exist. As a consequence, the number of reports concerning accidents in tunnels and other underground structures is relatively reduced. A review of general safety aspects in underground structures is analyzed in this lecture, namely different types of accidents and their consequences. The main causes of accidents and other deteriorations are described and typified, such as no adequate geotechnical studies, errors in the planning stage and in the calculations namely by numerical codes, errors during the construction or rehabilitation phase, and during the exploitation of the underground structures. Accidents and incidents in underground hydroelectric projects, in transportation tunnels and in urban subway systems are referred. The interaction of urban tunnels on the existing surface and subsurface infrastructures and the damage associated is also mentioned.

1 ACCIDENTS IN UNDERGROUND WORKS AND THEIR CONSEQUENCES

Most accidents and other associated problems occur during construction of geotechnical structures, and are very often related to uncertainties in side ground conditions. Therefore, it becomes essential to develop risk analysis systems and to avoid its occurrence. The associated risk has a complex nature resulting from the combination of two sets of factors, the events and the corresponding consequences, and the vulnerability factors that determine the probability of an event having certain consequences. Risk analysis illustrates the fact that decision-making must be based on a certain level of uncertainty (Einstein 2002; Caldeira 2002). Risk analysis is part of the decision-making cycle, as Figure 1 illustrates.

Uncertainty is a prevailing aspect in geotechnical engineering. Consequently, different categories may be established, namely: in the spatial variability and in the time of the geological factors; in errors introduced by measuring and in estimating geomechanical parameters, including statistical description; in model and load uncertainties; and, lastly, in omissions (Einstein 2002).

The use of underground space is increasing due to needs of sustained development. The occurrence of accidents or incidents in underground works is not as unusual as in other geotechnical structures. Nevertheless, the dissemination of those events is not common, in view of the legal and sociological problems involved. Usually, there is a tendency to minimize the dissemination of the involved hazards and their causes.

Up to relatively recently, risk assessment and risk analysis have not assumed particular relevance when evaluating underground projects and other major geotechnical projects. Nevertheless, the situation is changing and risk analysis has been successfully implemented, for major transportation infrastructures projects the USA and in Switzerland, using both commercial and research software for risk analysis. Special reference can be made to the system DAT – Decision Aids for Tunneling, developed by MIT (USA) in co-operation with EPFL – Ecole Polytechnique Fédérale de Lausanne

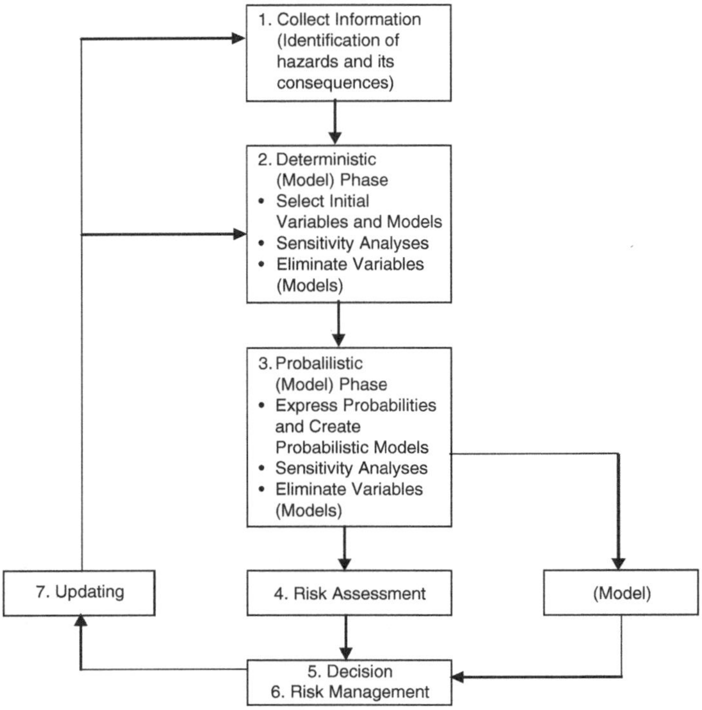

Figure 1. The decision analysis cycle (from Einstein 2002).

(Dudt et al. 2000). DAT is an interactive program using probabilistic modeling that permits to analyze the effect of geotechnical uncertainties and construction uncertainties on construction costs and time through probabilistic modeling (Einstein et al. 1999; Sousa et al. 2004).

The identification of the geotechnical risks aims to assess all causes that may lead to a hazard-inducing process. Therefore, this lecture intends to carry out a review study about the causes for accidents in underground structures using different construction methodologies, such as SEM (*Sequential Excavation Method*) that is usually designated as NATM, and also using tunneling boring machines. It also aims to study causes for accidents occurred during operation or due to exceptional actions. Reference is also made to damage associated with existing infra-structures on the surface or subsurface, due to the excavation of the works or to the tunnel itself, or due to several factors, such as aging.

The study of accidents in underground structures is a very important tool for understanding the instability phenomena and mechanisms that are produced by its construction and, as a result, it makes possible to select the most appropriate construction methods for future projects.

Even though the occurrence of accidents and incidents in underground works is not as unusual as in other geotechnical structures, due to the risks involved in their execution or due to natural disasters or fires, their dissemination is not always common. In fact, there is a tendency to minimize their dissemination and causes. This fact may lead to the repetition of former errors by designers and constructors. Therefore, the number of failure reported in underground works is comparatively reduced.

Recently, with the collapse, on 21 October 1994 and over the following days, of three parallel tunnels being constructed as part of the Heathrow Express Ray Link, at Heathrow airport, near London, the HSC (*Health and Safety Commission*) carried out an investigation intended to study the implications of the use of NATM (or SEM) tunnels, and the safety of the finished tunnel using this method when compared it with other tunneling construction methods. Furthermore, that investigation was also intended to determine the causes for the collapses occurred at Heathrow airport and to publish a report with the conclusions achieved (HSE 1996; ICE 1996).

Apart from published reports, a wider debate about safety of NATM tunnels in weak ground formations was promoted and to related tunneling using sprayed concrete linings and open face excavations.

The main conclusions achieved by the HSC Commission about the use of NATM tunnels were, among others, the following (HSE 1996):

- Major accidents in the excavation of NATM tunnels have occurred all over the world. Nevertheless, not always the most critical aspects as regards safety have been well understood by the technical community.
- Collapses occurred in NATM tunnels in urban areas may have serious consequences not only for the workers, but also for the surface infrastructures and for the environment. There are structural solutions, for which the collapse consequences are unacceptable, that is why, for these cases, alternative solutions must be found.
- The conducted geomechanical investigations conducted must ensure that there is no possibility of finding unexpected critical conditions regarding the safety of the underground structure. Consequently, a detailed accurate design becomes necessary and each structural element must be dully designed before its construction.
- An integrated procedure that will be able to consider the design of provisional and final supports must be developed. The design must take into account the whole procedure developed by the construction of tunnels according to the construction methodology followed.
- The tunnels excavated with supports designed in accordance with NATM method are as safe as those excavated using other construction methods.

A detailed analysis of the risk associated with the NATM method and its comparison with other methods has not yet been done. Nevertheless, each different methodology introduces its own hazards, which are highly dependent on the location and the function of the underground structure.

The excavation with tunneling boring machines (TBM), and particularly with EPB (Earth Pressure Balance) TBMs, is frequent because it is comparatively fast when compared with other solutions, being necessary to ensure the stability of the tunnel heading in front and to control accurately the surface settlements, in the case of urban tunnels (Babendererde 1999; Barton 2000; Vlasov et al. 2001).

In heterogeneous rock masses and for shallow urban tunnels, these aspects are difficult to accomplish. The geotechnical and hydrogeological ground investigations are always insufficient and cannot provide with detail and accuracy the strength and permeability characteristics, along the axis of the tunnel, as well as the side zones and in depth. The existence of buildings and other infrastructures does not permit to carry out boreholes and other survey actions on the surface and on all adequate places, being therefore necessary to operate at the excavation level. Consequently, boreholes are performed at the tunnel heading. Furthermore, the pressures and densities of the formations excavated in the excavation chamber, as well as the volumes of excavated materials are controlled with the highest possible accuracy, by establishing upper and lower limits in each excavation phase (Martins et al. 2003).

In the case of operating tunnels, accidents may also occur with partial or total damage of structures and equipment, as a result of fires, explosions and floodings, as well as during the rehabilitation works of old tunnels, as occurred in the Viérzy tunnel, France (Silva 2001).

Some major accidents have occurred in tunnels due to natural disasters, such as, avalanches, rock sliding, flooding and nonobservance with the safety requirements (Vlasov et al. 2001).

Other types of deterioration may occur (in the concept of deterioration are included the accidents, according to the methodology followed by ICOLD) in old and recent tunnels, which are mainly related with the rock mass and with the support. In the case of old tunnels, the anomalies are associated with a decompression around cavities, due to the use of the construction methods adopted at the time, which are particularly damaging for the underground structures. The anomalies in old tunnels are also related with deterioration of the support due to occurrence of voids, existence of void joints in the masonry, presence of water, wind erosion in the case of railway tunnels, and obviously, due to actions imposed by the rock mass onto the support (Freitas et al. 2003). In recent tunnels, the main deteriorations are essentially related with the construction method, namely NATM, excavation with TBMs, or by the cut-and-cover method. The supports are made of materials, like cast concrete, sprayed concrete with and without fibers, bolts, anchors and steel supports. Other malfunctions related with the design, optimistic considerations about drainage and consequently reduction of water pressures, errors in calculation and planning, may also justify the types of deterioration occurring rather frequently (HSE 1996; Matos 1999; Vlasov et al. 2001).

The number of accidents and incidents has significantly increased in the last few years for several reasons. These are mainly related with the growth in the construction of tunnels and with the fact that the associated risks are not dully known and controlled and, sometimes, with an excessive confidence on the construction methods. Since many accidents or incidents are not reported, it is not possible to define an adequate statistic about the main causes for occurrence of these situations. Therefore, it is only possible to carry out a description of the events occurred and to summarize the main causes for accidents or incidents occurred in underground works.

2 REVIEW OF ACCIDENTS AND MAIN CAUSES

The accidents during the excavation of tunnels are uncontrolled events that may have serious consequences. The frequency of accidental situations is relatively high compared with other geotechnical works.

As previously mentioned, HSE (1996) has carried out an extensive literature search in order to establish and analyze the incidents and accidents that have occurred in underground structures. A preliminary analysis of these data makes it possible to state as follows:

i) The number of collapses in urban areas is about two times higher than the ones in rural areas.
ii) The cases reported do not concern only the countries with reduced experience with the use of NATM.
iii) Most reported cases refer to railway or subway tunnels.
iv) The environmental consequences of collapses occurred (for public, infrastructures and buildings) are consistently high in urban areas.

Other cases of incidents and accidents have been reported, namely in Japan in 65 tunnels, mainly in hard rock (Inokuma et al. 1994). Among these cases, 15 refer in the range 50–500 m^3, and 3 were over 1000 m^3 of ground loss. There were two situations with craters on the surface.

A collection of accidents in tunnels in São Paulo was done by Neto and Kochen (2002). The majority of the reported cases were in soil formations and more reduced events in rock formations. As important conclusions to be drawn from those cases, reference is made to the fact that some accidents occurred in clayish variegated formations and in gray clays that were fairly cracked. The cracks lead to a reduced strength in these formations.

These incidents and accidents reported may be attributed to different factors, such as the fact that the NATM is used in increasingly difficult conditions in urban areas and the lack of knowledge by designers and constructors.

Nevertheless, only a reduced number of publications and technical papers about accidents in tunnels is available. Special reference is however made to the book published by Vlasov et al. (2001) and to the already mentioned article of Neto and Kochen (2002). Figure 2 shows a case of an accident occurred in São Paulo. Table 1 presents a summary of cases of accidents occurred in tunnels all over the world adapted from the publication of these two last authors.

In the case of NATM tunnels, most cases reported refer to collapses occurring close to the excavation front. These may be due to the collapse of the ground forming a hole above the tunnel, which, in the case of not very deep works, and particularly in the case of large span underground structures, may reach the surface and have disastrous consequences for the workers, for the public in general, for infra-structures and for the environment. Sometimes, the collapse of the front is

Figure 2. Tunnel collapse in São Paulo (Neto & Kochen 2000).

attributed to the instability conditions of the ground, when, actually the reason for collapse is the use of a construction method that is not appropriate for the site conditions.

Causes of collapse have been analyzed and the generalized collapse mechanisms can be grouped into three categories: i) ground collapse in the heading (Figures 3 and 4); ii) failure of the lining both before and after ring closure (Figure 5); and iii) other collapse locations and mechanisms.

Resuming, the various types of causes that have led to collapse are the following (HSE 1996):

i) *Collapse close to the excavation front*:
   - Collapse of unstable rock natural ground close to the excavation face
   - Collapse of the unstable excavated front face with inclusion of man-made feature such as borehole, well or culvert
   - Collapse of partly completed lining due to excessive settlement or convergence
   - Collapse of the bench in the longitudinal direction
   - Collapse of the bench during excavation towards the center of the tunnel
   - Collapse of the longitudinal "cantilever" of the heading in advance of the first section of completed ring
   - Collapse due to the crown excavation being too far in advance of the closure of the ring
   - Collapse due to failure of the temporary invert to the crown section
   - Collapse due to bearing failure under the "elephant's feet" to the crown section
   - Collapse due to structural failure of the partial completed lining – for instance, due to local overstressing or rock joint movements

ii) *Collapse in the area of the completed primary lining*:
   - Collapse due to excessive settlement or convergence
   - Collapse due to local overstressing, caused by unanticipated or unallowed for loading conditions

– Collapse as a result of substandard materials or significant construction defects
– Collapses due to interruption in the works concerning the junctions of old and new parts of the tunnel lining
– Collapse due to repairs, changes or corrections to the profile of the primary lining

iii) *Other collapse locations and mechanisms*:
– Collapses at portals, usually associated with weak geotechnical quality of the ground or loose rock or ground
– Collapses from vertical shafts, usually associated with weak geotechnical quality ground and (or) water on the outside of the shaft construction.

In the case of tunnels excavated using TBMs, the collapse close to the front may lead to severe damages and to the destruction of the cutting devices, which leads to additional access works and

Table 1. Accidents in tunnels (adapted from Neto & Kochen 2002).

Year	Place	Type of accident
1973	Paris	Railway tunnel (France), collapse
1981	São Paulo	Metro (Brazil), instability
1984	Landrücken	Tunnel (Germany), collapse
	Bochum	Metro (Germany), collapse
1985	Richthof	Tunnel (Germany), collapse
	Kaiserau	Tunnel (Germany), collapse
	Bochum	Metro (Germany), collapse
1986	Krieberg	Tunnel (Germany), collapse
1987	Munich	Metro (Germany), 5 collapses
	Weltkugel	Tunnel (Germany), *cave-in*
	Karawanken	Tunnel (Austria/Slovenia), large inflows and severe deformations
1988	Kehrenberg	Tunnel (Germany), serious surface settlements
	Michaels	Tunnel (Germany), collapse (pilot tunnel enlargement)
1989	Karawanken	Tunnel (Germany), collapse
	São Paulo	Metro Itaquera (Brazil), collapse
1991	Kwachon	Tunnel (Korea), collapse
	Seul	Metro (Korea), collapses affecting buildings and utilities
1992	Funagata	Tunnel (Japan), collapse
	Seul	Metro (Korea), 2 collapses
1993	Seul	Metro (Korea), 4 collapses
	Chungho	Taipei tunnel (Taiwan), collapse
	Tribunal da Justiça	Tunnel (Brazil), São Paulo collapse
	Toscana	Italy, severe deformations (collapse)
1994	Carvalho Pinto	Brazil, portal failure during construction
	Montemor	Road tunnel (Portugal), 2 collapses
	Galgenberg	Tunnel (Austria), collapse
	Munich	Metro (Germany), collapse
	Heathrow	Airport, London (UK), collapse
	Storebaelt	Denmark, fire in TBM
1995	Motorway tunnel	Turkey, collapse
1996	Motorway tunnel	Turkey, collapse
	Los Angeles	USA, collapse
	Athens	Metro (Greece), collapse
	Adler	Tunnel (Switzerland), collapse
	Toulon	Tunnel (France), collapse
	Eidsvoll	Norway, collapse
1997	Athens	Metro (Greece), collapse
	São Paulo	Metro (Brazil), collapse
	Carvalho Pinto	Metro (Brazil), collapse
1998	Russia	Collapse

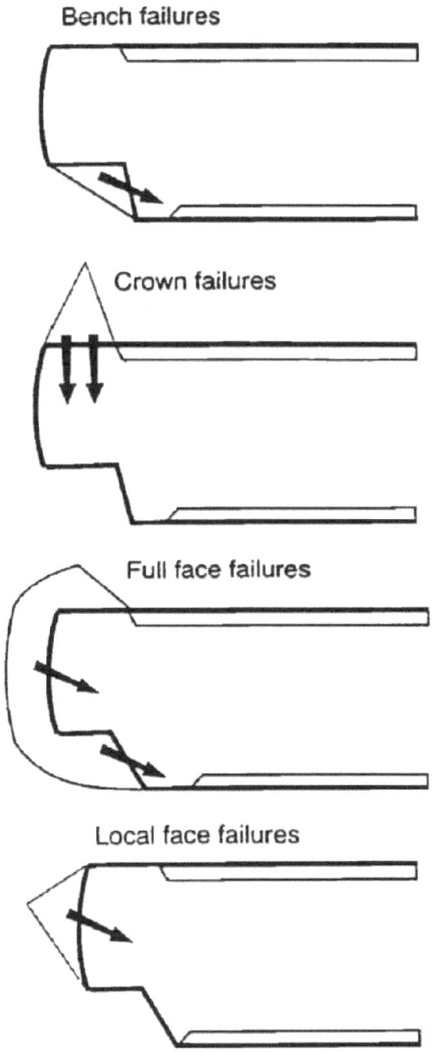

Figure 3. Collapse of tunnels at the excavation front in very low strength rock masses (HSE 1996).

therefore to significant costs and to extension of delay. In situ repairs of the TBM are usually extreme difficult or their dismounting and removal from the tunnel. A case history is illustrated in Figure 6. It shows a diagram of the works performed in a tunnel referring to sewage systems in Canada, with about 45 km, in the vicinity of Montreal. Figure 7 shows different cases of collapse with the use of TBMs.

Lastly, in order to revise the causes that have led to collapse, various main types can be -considered (HSE 1996; Vlasov et al. 2001):

i) *Unpredicted geotechnical conditions*
It is associated with non-rigorous geotechnical characterization caused by a reduction in the geotechnical survey works. A common unpredicted situation is related with washing or erosion of the ground. Therefore, it is good practice to continue the geotechnical survey during

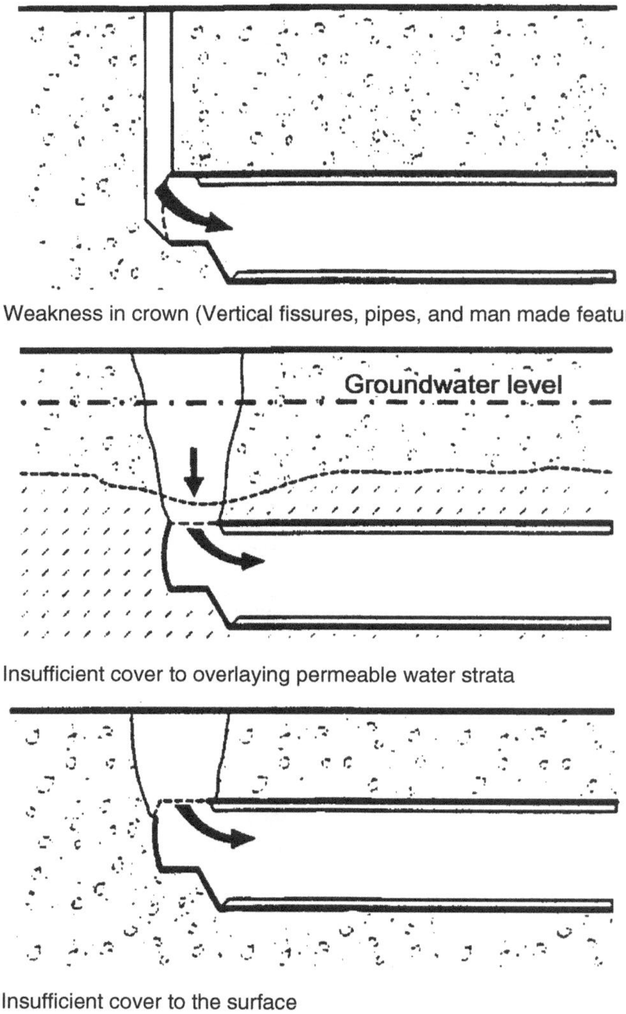

Figure 4. Collapse of tunnels at the excavation front due to specific conditions (HSE 1996).

construction and to carry out an analysis of the excavation front by experts. This is one of the causes that is most often reported in many collapse situations.

ii) *Errors in design and in specifications*

Tunnel failures have occurred due to failure at the planning stage to locate underground structures such as shafts, culverts and unfilled or poorly filled boreholes. Other causes mentioned are related with the existence of an inadequate ground cover the tunnel, with the excavation and support measures specified without taking into account the type of formations involved, with the use of a classification system that has led to non-adequate supports, with inadequate specification of the construction materials and lastly to inadequate planning for the unexpected or emergencies.

iii) *Errors in calculation or numerical analyses*

These include errors of calculation during design and frequently in connection with monitoring, the latter being related with the quality of the observation data. Other situations of reported errors to be considered are related with the adoption of inappropriate calculation parameters,

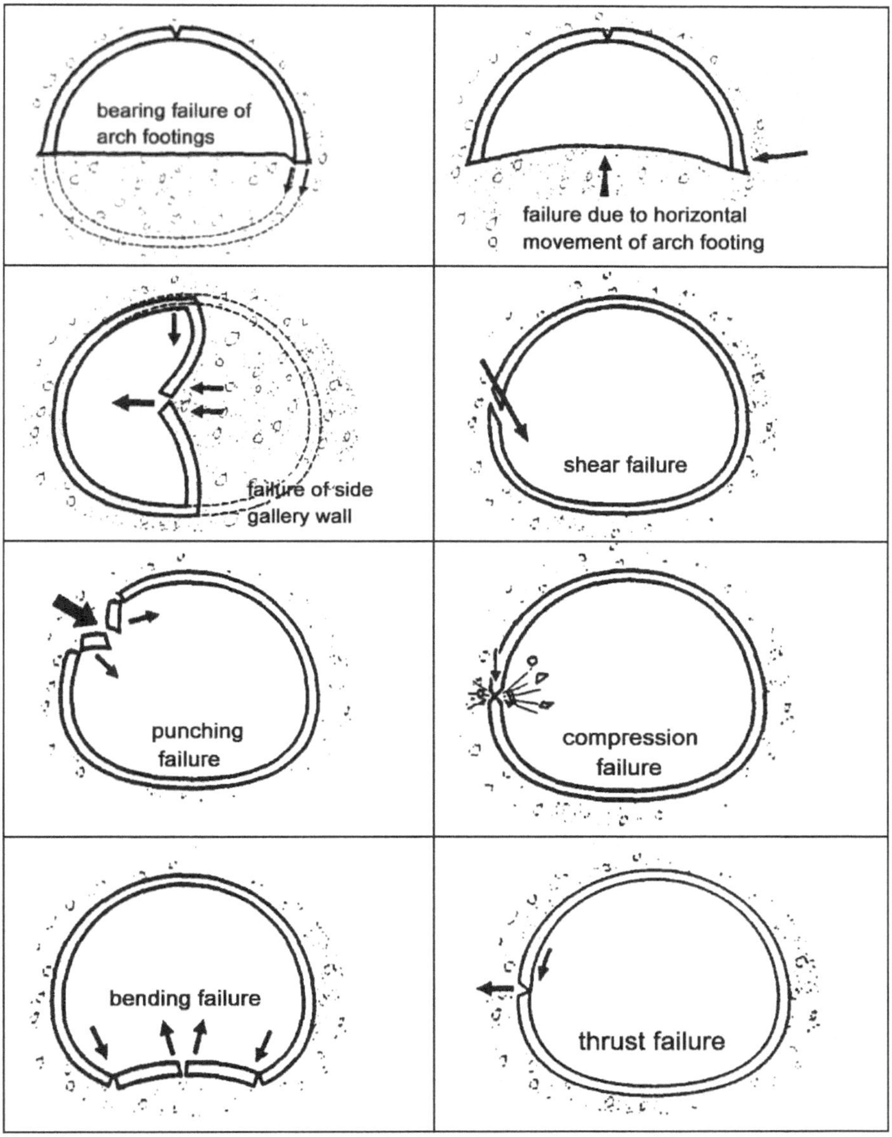

Figure 5. Failure mechanisms of lining before and after ring closure (HSE 1996).

with underestimation of the effect of water and sometimes with the use of inadequate or non-validated calculation programs.

iv) *Errors during construction*

Construction errors are extensive and difficult to specify. The most common are the following:

- Lining without the specified thickness
- Faulty installation of rock anchors and arches
- Inadequate installation of ground freezing pipes
- Incorporation of excavated material in the invert concrete
- Inadequate profile of the invert and badly executed lining repairs.

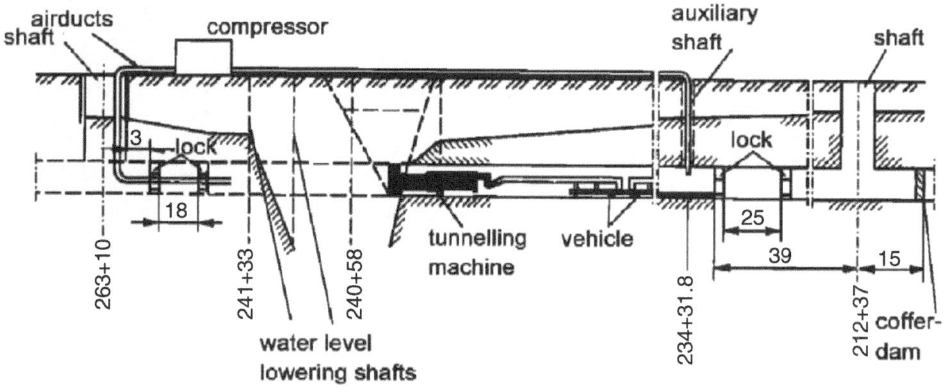

Figure 6. Works performed for an emergency situation due to collapse at the tunnel heading (Vlasov et al. 2001).

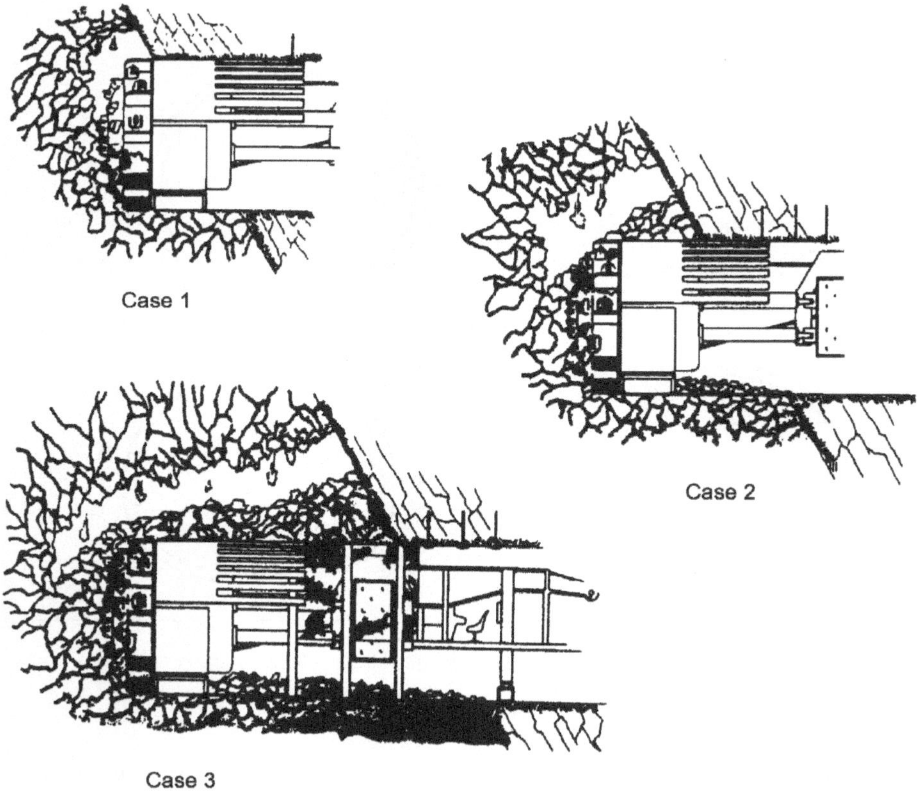

Figure 7. Failure cases for rock TBM tunnels (Barton 2000).

v) *Errors in the control and management*
These include, among others: inexperienced designers and contractors; lack of adequate conclusions after the occurrence of situations that indicated the existence of a poorly adequate structural design; poor site inspection; and, lastly, adoption of an inadequate construction sequence.

To summarize, factors responsible for accidents in tunnels are mainly natural and/or technological. The natural factors involve essentially features such as: the structure and properties of the formations, heterogeneity, groundwater conditions, geological and physical processes, including seismic events, karst erosion and geothermal. The technological factors are related with the man' engineering activities. These are for instance, the perturbation of the *in situ* state of stress and the deformational movement induced by the excavations, the interaction with the surface infra-structures and with those inside the very rock mass, the decrease and increase in the phreatic levels, the disregard for the construction criteria, as well as the operational conditions for a tunnel in service (Vlasov et al. 2001).

3 UNDERGROUND HYDROELECTRIC SCHEMES

An underground hydroelectric power scheme comprises a variety of works, being usually, a dam, an intake, the hydraulic circuit which can be grouped into high and low pressure circuits, access tunnel and shaft, surge chambers, the powerhouse complex, with electromechanical equipments, transformers and other complementary equipment (Sousa et al. 1994).

The use of an underground arrangement for the powerhouse has significant advantages from a point of view of safety and environment. It may consist of a set of large underground works, assuming a decisive part in the design and construction of the works. The safety criteria commonly used consist of limiting the displacements, which indirectly restrains deformations. Other criteria can also be established for the supports, by considering safety stresses or by being dimensioned having the failure as reference.

One of the main requirements for a successful excavation is to have an accurate knowledge of the specific geological structures, such as the occurrence of low strength surfaces. Accidents or failures may occur due to block falls, planar or wedge failures and to the use of inadequate supports. The existence of discontinuities with clay fillings is a situation to be taken into account in the failure process. The support system may comprise the removal of the clay from discontinuities and its backfilling with concrete, as well as for wide zones by the use of grouted bolts or anchors.

Case history situations in which accidents have occurred by planar or wedge failures are reported by Hansen (1993) for underground hydroelectric schemes in Sweden. Figure 8 shows

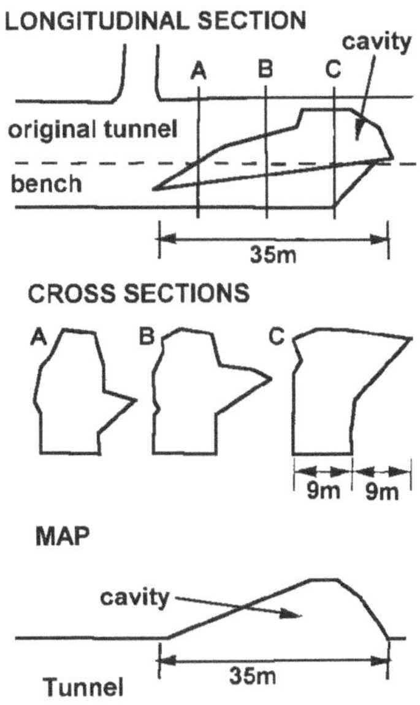

Figure 8. Planar failure in the Höljebro hydroelectric powerhouse (Hansen 1993).

the occurrence of a planar failure in the Höljebro powerscheme. The first unit at Höljebro was taken into operation in 1932. In 1978, a power reinforcement having been planned. The section of the tailrace tunnel increased from 55 m^2 to 110 m^2. During excavation works, a planar failure occurred along 35 m length of the tunnel, as Figure 8 illustrates.

As mentioned before, one of the main causes for accidents in large caverns associated to underground hydroelectric schemes, results from the occurrence of low strength continuous surfaces, conjugated with the usual discontinuity sets occurring in the rock masses.

A case that had significant consequences, reported by Rocha (1977), is an accident occurred during construction of one of the surge chambers of the Cahora-Bassa hydroelectric system, in Mozambique. The underground powerhouse complex that includes the surge chambers is located at a depth ranging from about 130 to 230 m. The specific dimensions of the powerhouse cavern correspond to a length of 216.7 m, a width of 28.9 m and to minimum and maximum heights of 24 and 57 m, respectively. The two surge chambers, set up in parallel with the powerhouse, have the following specific dimensions: a length of 82.5 and 87.7 m for the north and south surge chambers, respectively; a width of 19.0 m for both of them; and a height of 72.0 and 70.3 m for the north and south chambers, respectively (Silva 1988; Sousa et al. 1995). Figure 9 shows a schematic view of the underground powerhouse complex.

The accident occurred in one of the surge chambers, as Figure 10 illustrates. It consisted of a wedge failure, with a volume of about 2000 m^3, which is schematically represented in Figure 11 (Rocha 1977).

In the site where the hydroelectric scheme was set up, there are formations of the higher Precambrian essentially composed of granitic rocks with zones of reduced importance of gabbrodioritic rocks and granulites, crossed by aplite, porphyry and pegmatite veins and by lamprophyric and gabbroric beds (Sousa et al. 1995). In the region, three tectonic stages may had occurred responsible by the discontinuity system of the rock mass, some faults of reduced geotechnical importance and three major families of discontinuities, one being sub-horizontal and two of which were inclined (Silva 1988). The sub-horizontal lamprophyre dykes that intersect the surge chamber are accompanied in the ceiling and wall by gneiss formations. The values as follows were obtained for the mean strength of the discontinuities (Silva 1988): gneissic granite – $\phi = 41°$ and c = 0.29 MPa; lamprophyre – $\phi = 20.3°$ and c = 0.22 MPa.

The accident was due to a wedge failure that took place along the intersection line of the two inclined discontinuity plans belonging to the family of discontinuities, in view of the occurrence of a low strength surface with a very low friction angle joints. The discussion of these situations is thoroughly analyzed in the publication of Hoek and Bray (1973).

The pressure tunnels and shafts are elements of the hydraulic circuit of considerable importance and complexity in hydroelectric and hydraulic schemes. They place special problems, in which the

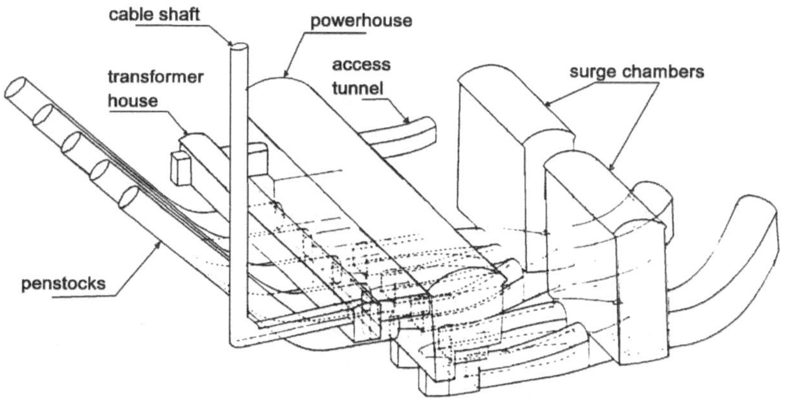

Figure 9. Cahora-Bassa underground powerhouse complex.

Figure 10. Accident occurred in the surge chamber (Rocha 1977).

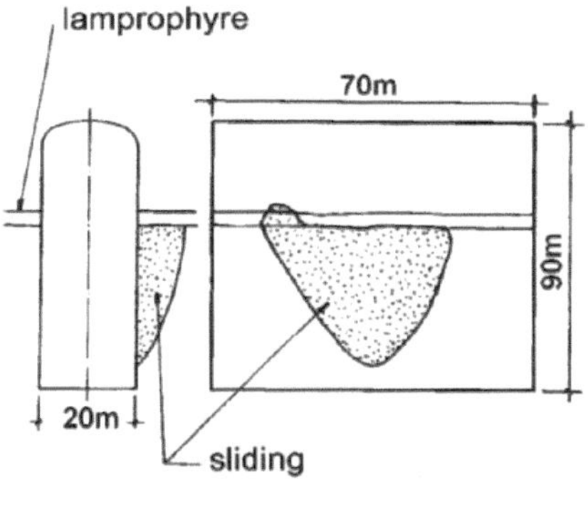

Figure 11. Scheme of the accident in the surge chamber (adapted from Rocha 1977).

hydromechanical behavior of the rock masses is relevant (Lamas 1993; Silvestre et al. 2002). These are usually carried out on a good quality rock mass, in which the rock mass represents a significant contribution to the strength to high internal pressures.

Several cases of accidents and other damages (deterioration according to ICOLD) in pressure tunnels have been reported by several authors (Brekke and Ripley 1987; Lamas 1993; Hoek 2000). Table 2 shows different classes of deterioration which correspond to typical situations occurring in these works.

The majority of deteriorations occurred in tunnels with concrete supports or without any support. The first infilling or the preliminary tests were responsible of about 20% of the cases. However most cases occurred during the operation of the hydroelectric schemes. According to a study conducted by Lamas (1993), Figure 12 shows the distribution of the deterioration cases, in percentage, for the situations of unlined and concrete pressure tunnels. For steel lining high

Table 2. Classes of deterioration in pressure tunnels (Lamas 1993).

Class	Description of the deterioration
A	Inadequate confinement, leading to excessive rates of flow, hydraulic jacking or instability of the rock mass, including landslides or uplift
B	Specific geologic features of high hydraulic conductivity, leading to leakage, hydraulic jacking or instability of the rock mass, including landslides or uplift
C	Deterioration of the rock mass, namely due to erosion of seams, dissolution and swelling, leading to excessive leakage, rockfalls or rock mass instability
D	Excessive water pressure as regards impermeable barriers, such as seams or clay filled faults, leading to rock mass movement and instability, including landslides
E	Deformable rock mass, inefficient grouting or deficient construction, leading to failure of the lining, namely due to internal water pressure
F	Buckling of steel linings caused by external pressure of water or grouting
G	Dynamic fluctuations of water pressure

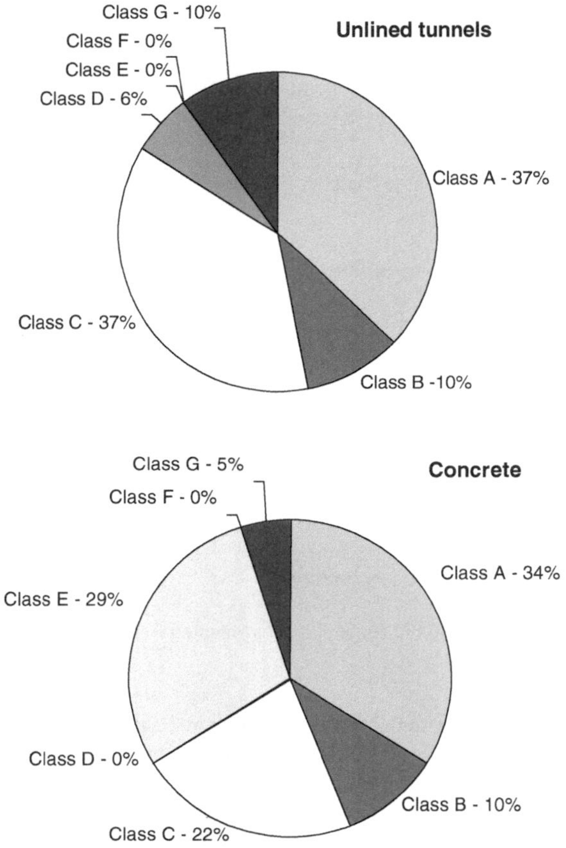

Figure 12. Distribution of the classes of deterioration for unlined and concrete pressure tunnels.

pressure tunnels the cases of deterioration studied are mainly included in classes E and F, with only few situations in classes D and G. The analysis of the figure makes it possible to identify the most important mechanisms, which are responsible for deterioration of the pressure shaft and tunnels.

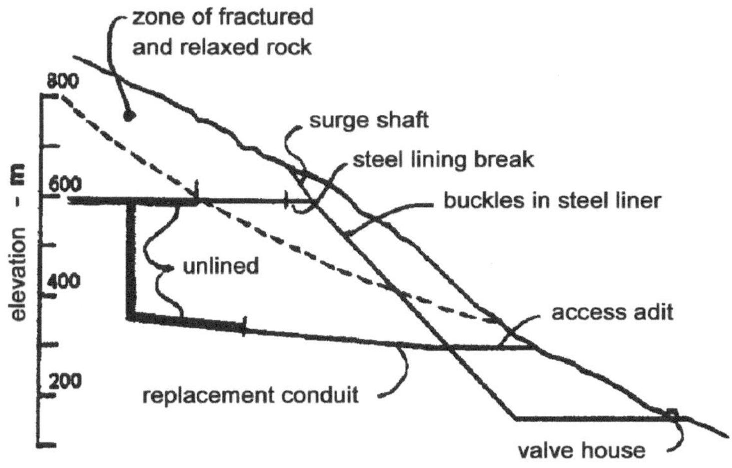

Figure 13. Cross section of a high pressure tunnel of the Wahleach hydroelectric scheme (Hoek 2000).

In the case of pressure tunnels locate in slopes, severe accidents may occur, as in the case of the Wahleach hydroelectric scheme, in British Columbia, Canada (Figure 13). A break in the steel lining in this scheme occurred in January 1989 and it is thought this break was caused by a slow down-slope gravitational movement caused by block rotations within a near-surface zone of loosened jointed rock (Hoek 2000). The Wahleach project is located 120 km east of Vancouver. Water flows through a 3500 m long 3 m diameter unlined upper tunnel, a rock trap, a 600 m long 2 m diameter concrete encased steel lined shaft inclined at 48° to the horizontal, a 300 m long lower tunnel and a 485 m long surface penstock to the powerhouse. The tunnels were excavated mainly in granodiorite which varies from highly fractured and moderately weathered. Two main joint sets occur in the rock mass, one set striking parallel to the slope and the other perpendicular to it. Both dip very steeply. Average joint spacing range from 0.5 to 1 m.

The water conduit operated without incident between the initial filling until May 1981 when leakage was first noted from the upper access adit located near the intersection of the inclined shaft and the upper tunnel. This leakage stopped when two drain pipes embedded in the concrete backfill beneath the steel lining were plugged at their upstream ends. Large holes had been eroded in these drainage pipes where they were not encased in concrete and it was concluded that this corrosion was responsible for the leakage. This conclusion appeared to be valid until 25 January, 1989 when a much larger water flow occurred. As a result of the failure of the steel lining, rehabilitation measures were adopted (Figure 13) and described in detail in the publication of Hoek (2000).

The generation of cavities by dissolution in tunnels of hydroelectric schemes, as well as in railway tunnels, is a situation that may lead to serious cases of deterioration in tunnels, with effects on the safety. Gysel (2002) published an article that describes the processes of dissolution in hydroelectric tunnels, presenting two case histories in Switzerland and in Guatemala. Figure 14 presents a section of a pipe of the Engadin hydroelectric system, Switzerland. The figure shows the configuration of the cavities obtained by karstification of the rock mass, as well as the reinforcement used with concrete filling and bolts.

4 UNDERGROUND WORKS IN TRANSPORTATION SYSTEMS

The use of underground space for transportation systems began in Europe and in North America just before the first half of the XIX century, with the beginning of railway transportation. In the

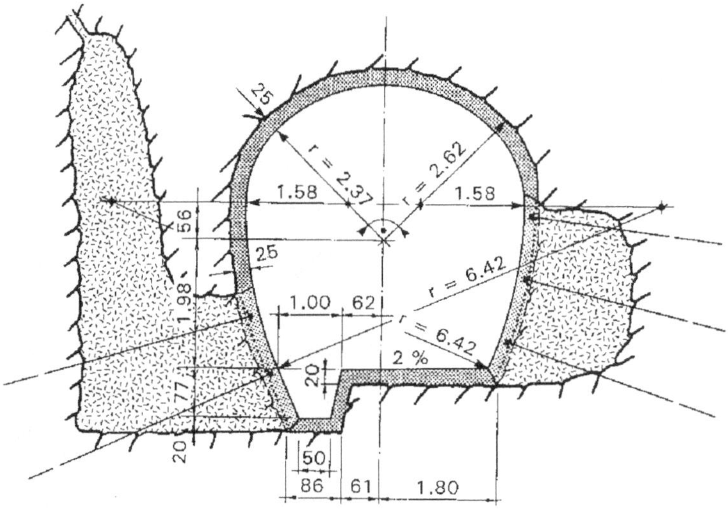

Figure 14. Engadin hydroelectric system. Cavities by dissolution (Gysel 2002).

last few years, many large underground projects are under construction in Europe and in the world. As example is the construction of new railway transport system (TGV) for all European countries including Portugal. Big projects in planning stage include an underground pneumatic transport system connecting major cities in Switzerland.

Different types of accidents in tunnels and in other cavities and the main causes of accidents were analyzed in the previous section. Nevertheless, the complex problems that are to be solved in this type of works are those associated with the execution of ventilation systems, but especially those resulting from major accidents caused by fire, as was the case of Mont Blanc and Tauern tunnels in 1999 (Sousa 2000).

The first world largest Seikan railway tunnel, which links the two major Japanese Islands, Honshu and Hokkaydo was under construction for 21 years (1964–1985). It is 54 km long, 23 km being under the bottom of the Tsugaru strait, 100 m deep under the bottom and 240 m maximum depth from the water surface. Figure 15 shows a longitudinal profile and a cross-section.

The tunnel was constructed simultaneously in 9 sections under complex geotechnical conditions. In the northern part it crosses sedimentary rocks, in the central part argillites and aleurites, and in the southern part cracked andesites, tuffs and breccias (Vlasov et al. 2001). The major part of the tunnel crossed heavily fissured rocks, with 9 large shear zones. Besides the main railway pilot and service tunnels were driven as well as shafts.

Despite the hazard prevention measures adopted during construction, 4 major accidents related to tunnel flooding took place. The first water inflow occurred in February 1974, in an inclined shaft at a distance of 1 223 m from Honshu Island. As a result half of the excavation was filled with water. The measures adopted to overcome the consequences of the accident comprised waterproof dams and the restoring of the damage section. The second accident took place January 1974 in the crown of the service tunnel drives through disturbed igneous rocks.

The fourth and the most severe accident occurred in May 1976, during the construction of the service tunnel, at a distance of about 4.5 km from Hokkaydo Island. The maximum water inflow was 70 m^3/min under a maximum pressure of 2.8 MPa, exceeding the capabilities of the water removal equipment. In the first 3 days 120 000 m^3 of water flooded 3015 m of the service tunnel and 1493 m of the main tunnel. In 19 days the tunnel was drained and in addition a 290 m by-pass adit has been driven at distance of 60 m from the previous route.

Figure 16 tries to show the consequences of the accidents occurred in the Seikan tunnel. First section of the Figure shows a longitudinal section at a distance of 3509 m from Hokkaido island,

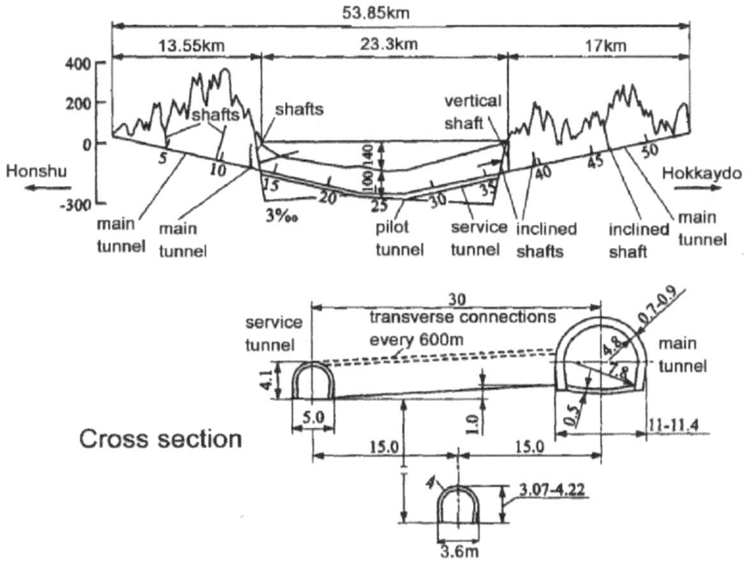

Figure 15. Seikan tunnel.

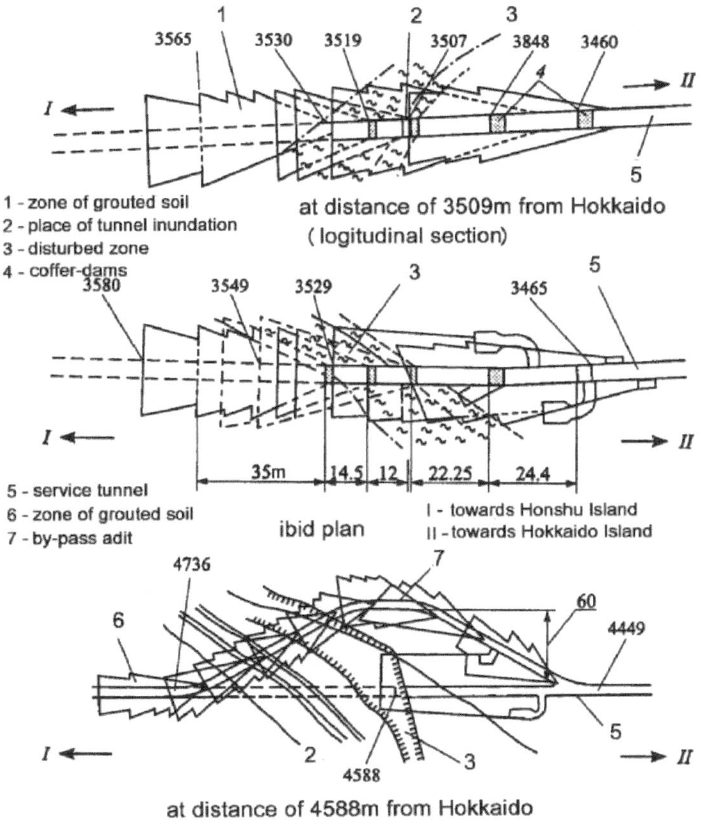

Figure 16. Consequences of the flooding of the Seikan service tunnel (Vlasov et al. 2001).

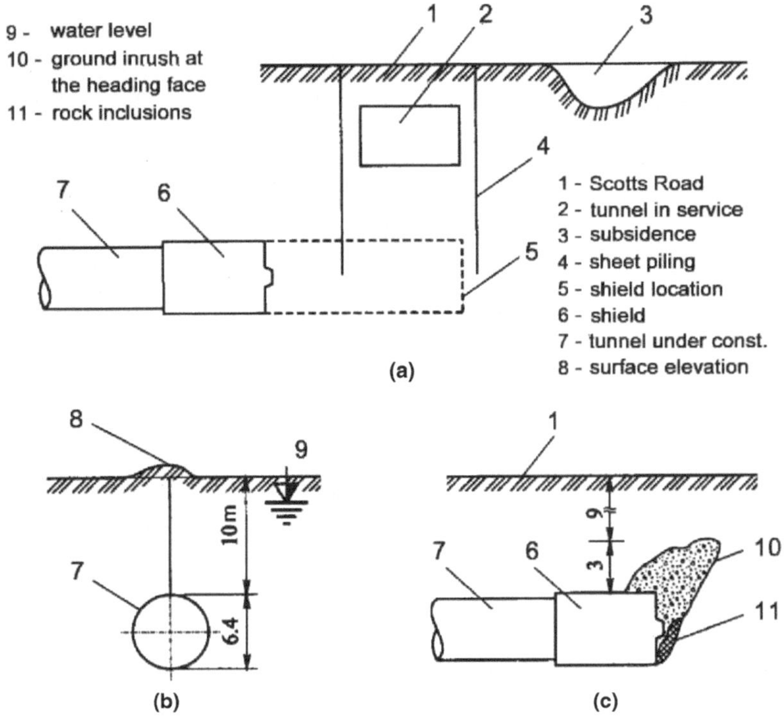

Figure 17. Surface deformations and fall of blocks during the construction of a tunnel of the Singapore metro (Vlasov et al. 2001).

second section shows a plan of the same location, and last section a profile at the distance of about 4588 m.

Regarding construction in urban environment and the use of TBMs, reference is made to the experience of construction of metro tunnels in Singapore, where several accidents occurred caused by rock falls in the heading. Figure 17 shows surface deformations and ground falls during excavations Orchard and Newton stations. More elements about these accidents may be consulted at the publication of Vlasov et al. (2001).

Grasso et al. (2003) presents a study case of an excavation of a tunnel with a TBM in heterogeneous granite formations, where problems occurred.

Earthquakes rarely introduce significant damage in underground structures. However, the 1995 earthquake, in Kobe, Japan, caused significant damages in few underground works. The earthquake occurred on January 17, 1995 in Kobe and Osaka regions. The moment magnitude of the earthquakes was 6.9 and strong motion lasted about 20 seconds, having caused extensive damages (Uenishi and Sakurai 2002). The worst damage to the underground facilities of Kobe was the collapse of the Daikai underground station built by the cut-and-cover method (Figure 18). The earthquake also caused damages in the Bantaki tunnel, in the mountainous region surrounding Kobe (Figure 19).

Figure 20 aims to represent the main damage occurred in the Daikai station. A more detailed explanation of the possible reasons of the failures occurred in these underground works is presented in the paper of Uenishi and Sakurai (2002).

In the particular case of cut-and-cover tunnels, problematic situations are likely to occur due to several factors, including:

i) Non-adequate calculation models, namely regarding the discretization used in the reinforced concrete support elements, the disregard for the various construction sections that are decisive for the design and the evaluation of generalized forces induced on the support elements.

ii) Inadequate geotechnical characterization of the different formations of the foundations and of their bearing capacity.
iii) Adjacent slopes movements responsible for deformations imposed on the reinforced concrete structure.
iv) Disregard for the seismic actions that may lead to an increase in the loads transmitted to the foundations of the concrete support structure.
v) Occurrence of sliding on the provisional slopes.

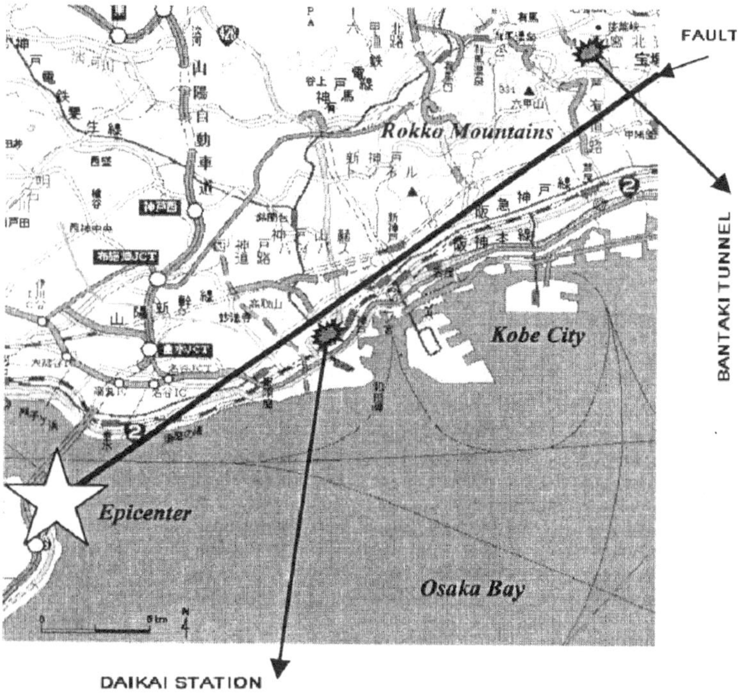

Figure 18. Failures in two tunnels of Kobe area (Uenishi & Sakurai 2002).

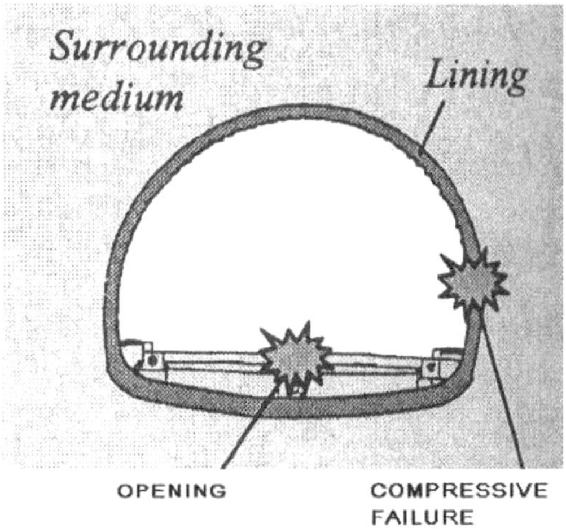

Figure 19. Failures in the Bantaki tunnel (Uenishi & Sakurai 2002).

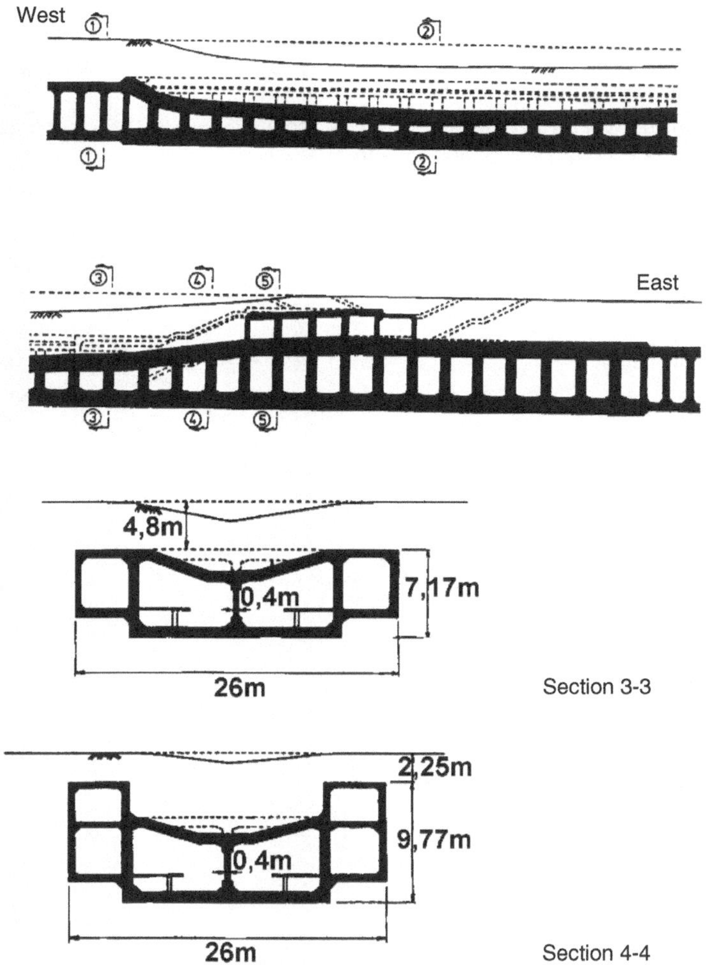

Figure 20. Failures in the Daikai station (Uenishi & Sakurai 2002).

An important cut-and-cover tunnel in Portugal due to its dimensions is the Grilo tunnel about 350 m long that belongs to the Lisbon internal regional circular in the branch Olivais Basto-Sacavém (Figure 21), (Barradas et al. 2002; Lemos et al. 2003). The adopted solution for the tunnel has a reinforced concrete double arch, a continuous transversal section and direct foundations. Most of the provisional slopes that had to be excavated had a high longitudinal involvement (Figure 22). Final excavation slopes were also cut, with a significant height, at the zone of the west mouth of the tunnel. The geotechnical formations involved consisted of limestone and soil layers, like sand and clay.

The behavior observed on the excavation of the provisional slopes basically presented two types of anomalies (Barradas et al. 2002): i) occurrence of distortional movements until great depths; and ii) surface sliding that affected mainly the north slope.

An example of occurrence of a problematic situation after the construction is the case of the 250 m long cut-and-cover tunnel with two arches, which is located on the Motorway A15 that links Navarra to Guipúzcoa, in Spain (Ortega 2003). The construction method consisted of excavation on rock, with about 30 m maximum height, on the left side, and with about 15m on

Figure 21. Grilo tunnel.

Figure 22. Slopes of the Grilo tunnel (Barradas et al. 2002).

the right side, of the execution of the support structure with two arches and of subsequent cover so as to reconstitute the original topography (Figure 23).

The tunnel was designed with pre-fabricated reinforced concrete elements, each arch being 2.3 m long and 35 cm thick. Due to nonsymmetrical forces applied on the reinforced concrete structure, movements and cracks were detected on the arches of the tunnel, which has immediately led to a reinforcement using steel linings, while other possible solutions for reinforcement and rehabilitation of the structure were being studied.

The solution chosen was to carry out with descendant anchors, with an inclination of about 5°, which were placed on the highest possible position. That solution was complemented by the execution of sub-horizontal drains drilled on the bottom part of the tunnel (Figure 23).

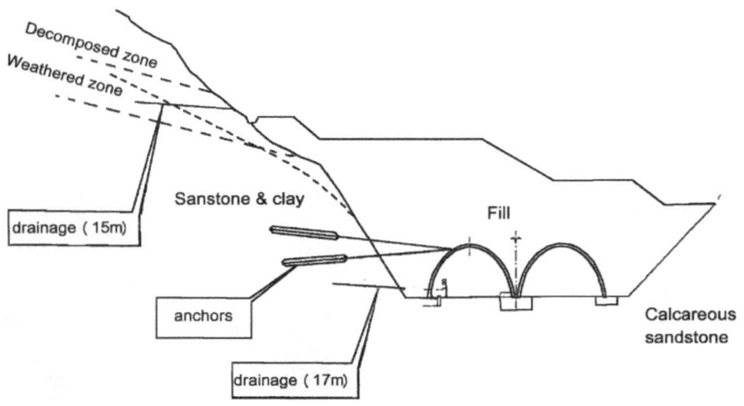

Figure 23. Tunnel on the Leizarán motorway (adapted from Ortega 2003).

5 INTERACTION WITH SURFACE INFRA-STRUCTURES

The construction of tunnels in urban areas inevitably results in movement of the ground around them. It may have a significant environmental impact, due to the possible occurrence of accidents or important damage on the infrastructures on the surface or on the subsurface, as well as due to noise and vibration, especially during the construction process. Therefore, it becomes essential, from a point of view of design and planning, to develop rational methods aiming to minimize the associated hazard or damage (Sousa 1998; Burland et al. 2002).

One of the important requirements is related with the assessment of the bearing capacity of buildings in the vicinity of the tunnel, since these structures are sensitive to settlements, to horizontal strains and to differential displacements. The quantification of the damages in the buildings is a highly subjective issue, which can be affected by various factors, such as the field experience, the approaches adopted by the design engineers and by insurance companies. Since there is the possibility of occurrence of damages due to movements induced on the surface, these must be subject to classification. The types of damages in buildings can have the classification as follows: i) visual or aesthetic; ii) functional; and iii) stability-related. In visual terms, 6 damage categories are defined, ranging from 0 (minor) to 5 (highly significant) (Burland et al. 2002).

There are various criteria that relate the values of settlements of a zone with the damages caused in buildings. A group of criteria relates the damages in the works with the maximum deflection of the deformation occurred in the foundation of the structure. However, other criteria relate the damages with the occurrence of movements in the ground, assuming that these result from the maximum strain developed in the structural walls, which depends not only on distortion, but also on horizontal deformations (Boscardin and Cording 1989).

Burland established a damage criterion, by a relationship of damage category to deflection ratio and horizontal tensile strain. That approach is illustrated in Figure 24, in which the building is represented by a rectangular beam of length L and height H. Different solutions are presented depending on either the structure is located on a sagging or hogging zone. Figure 24 shows two extreme modes, bending only about a neutral axis at the lower fibre and shearing only. In the case of bending, the maximum tensile strain occurs in the upper fibre and that is where cracking will initiate. In the case of shear, the maximum tensile strain are inclined at 45° giving rise to diagonal cracking.

The movements induced on the surface involve not only sagging and hogging profiles, but also significant horizontal strains. As an alternative to calculating tensile strains, charts are available, like in Figure 25, relating damage categories directly to deflection ratio and to horizontal strain. Figure 25 is related to L/H = 1 and for hogging mode. In order to assess the evaluation of the

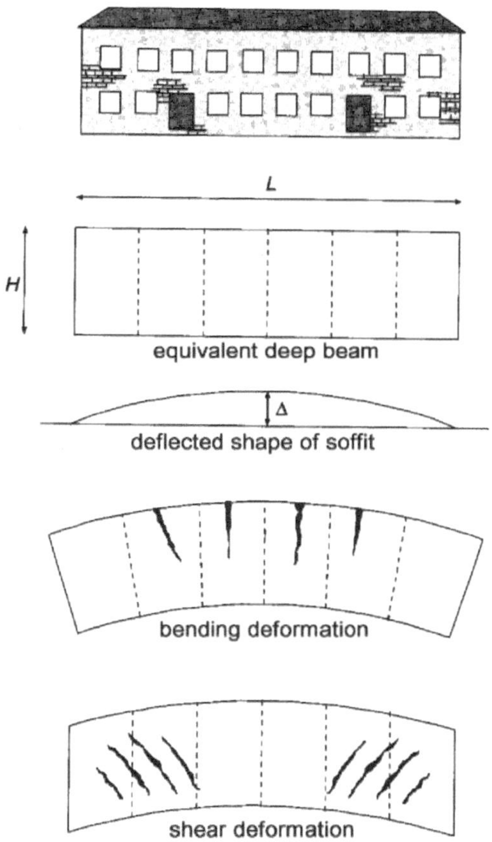

Figure 24. Representation of a building by an equivalent beam to illustrate cracking in bending and in shear (Cost C7 2003).

effects induced by the tunnel on the surface buildings, it is not usual to consider their stiffness. Nevertheless, the approach developed by Potts and Addenbrooke (Cost C7 2003) involves the use of design curves that modify the damage parameters conventionally calculated. Boscardin and Cording (1989) developed the concept of different levels of tensile strain using case records of damage cause by subsidence on a variety of buildings. They showed that categories of damage could be related to ranges of tensile strain as indicated in Table 3.

Figure 26 schematically represents three modes of building movements for the transverse settlement. The longitudinal settlements produced close to the excavation front and behind the front may also bring damage to buildings, but is more difficult to quantify.

The excavation of tunnels may also affect the subsurface structures, such as existing tunnels and service installations, namely sewers, gas mains, electric cables and pipelines (Figure 27).

In terms of serviceability state, effects of long-term groundwater flow and consolidation should be considered. The new tunnels may act as a drain, causing a permanent lowering of the phreatic level and consequent surface settlement on the ground surface above, as Figure 28 shows. It may affect other buildings outside the initial zone of influence of the tunnel.

The cases of instability of tunnels, which have been analyzed in detail, in the first chapters, cannot be accurately predicted, being possible to use for the purpose the analyses of observation results, namely the alarm and warning criteria (Sousa 2001). The instability of the tunnel may be transmitted to the surface. Figure 29 shows two situations, one with insufficient resistance from

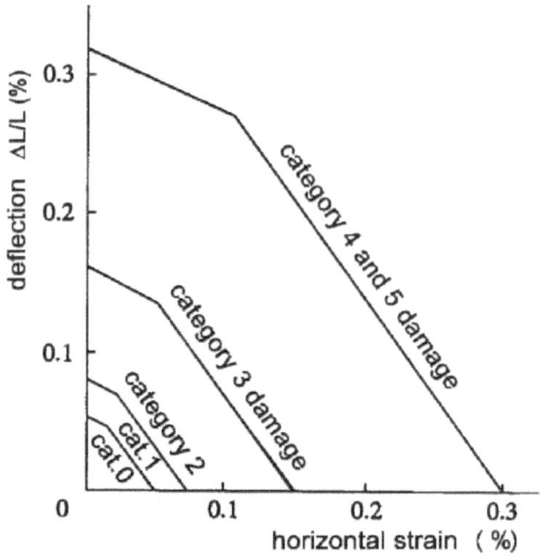

Figure 25. Damage criterion proposed by Burland for buildings located on the hogging part (Burland et al. 2002).

Table 3. Relationship between category of damage and limiting tensile strain after Boscardin and Cording (1989).

Category of damage	Degree of severity	Limiting tensile strain ($\varepsilon_{\mathrm{lim}}$) (%)
0	Negligible	0–0.05
1	Very slight	0.05–0.075
2	Slight	0.075–0.15
3	Moderate	0.15–0.3
4 to 5	Severe to very severe	> 0.3

the abutments to maintain stability of the lining and another, where insufficient face pressure has been maintained.

In many of the illustrated cases, it was possible to identify possible reasons for resulting excessive ground movement and damage. Establishing the causes allows counteract methods to be implemented, which are related to a good construction quality and protective measures. The causes of excessive ground movement are mainly related to inadequate initial investigations, poor design and analysis and poor control of construction works (Cost C7 2003). A number of protective measures are available. The in-tunnel measures include actions taken inside the tunnel during its construction in order to reduce the magnitude of displacements. Improvement in the TBMs had led to greater control of ground movements through controlling pressures at the face and grouting behind the lining. In the case of open-face tunneling, typical measures can be related to tunneling method, face support measures, excavation by parts, pilot tunnels, mechanical pre-cutting. Ground treatment measures are also adopted in order to reduce and modify the ground movements. Methods of permeation grouting and compensation grouting have become popular in recent tunneling projects. Structural measures are yet adopted by increasing the capacity of the structures, by increasing the ability of the foundations in order to resist the predicted movements, by stiffening the structure in order to modify the predicted movement, by making the structure less sensitive or by controlling the building movements by isolating it from the foundation.

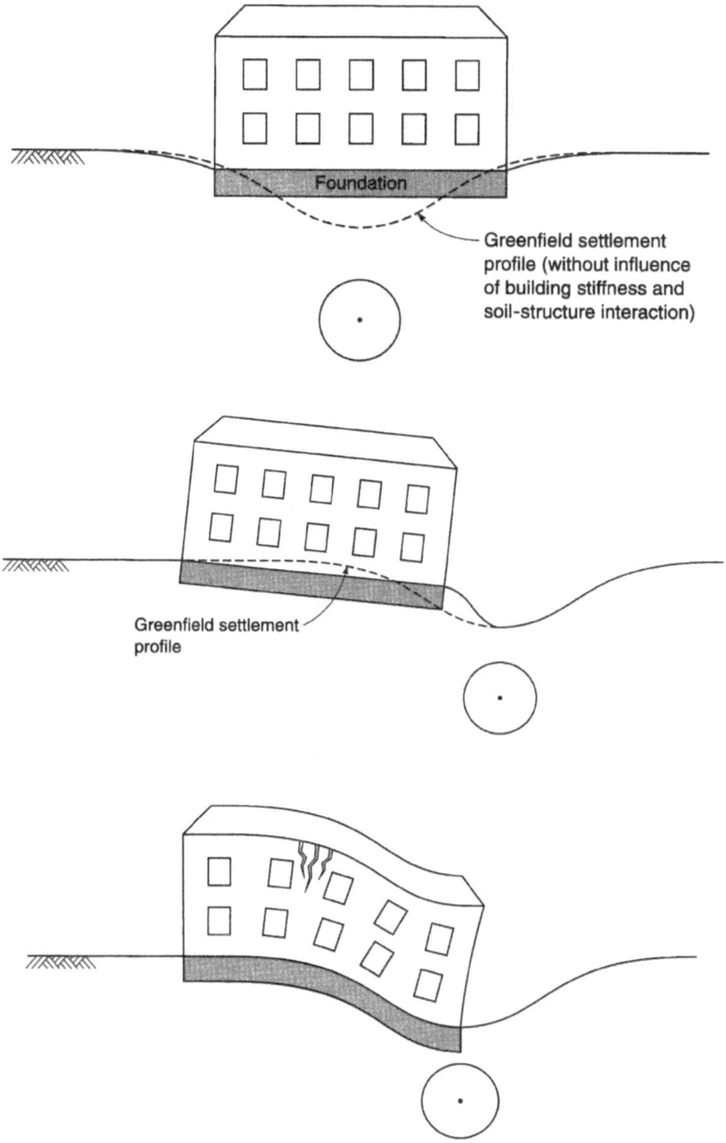

Figure 26. Representation of a building by an equivalent beam to illustrate cracking in bending and in shear (Cost C7 2003).

Finally, a very interested case history is presented referring to the Baixa-Chiado station of the Lisbon Metro, where it was necessary to reinforce the rock mass by compensation grouting so as to minimize the damage for the buildings in the neighbourhood (Barreto et al. 1999).

The *Baixa-Chiado* station was developed beneath the historical area of *Chiado* that was classified by UNESCO as patrimony of mankind. It comprises a considerable group of buildings that presented some anomalies resulting from the alterations occurred during their lifetime. That station started operation recently and serves two metro lines. Due to its strategic location it is presently one of the stations with highest passenger traffic.

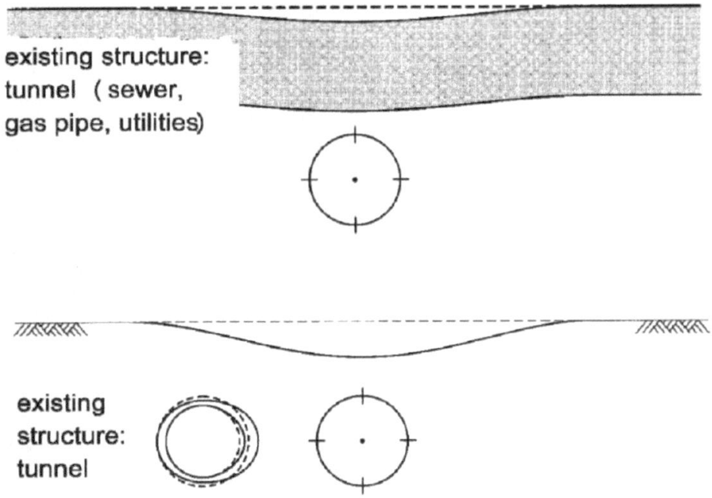

Figure 27. Influence of transverse settlement on subsurface structures (Cost C7 2003).

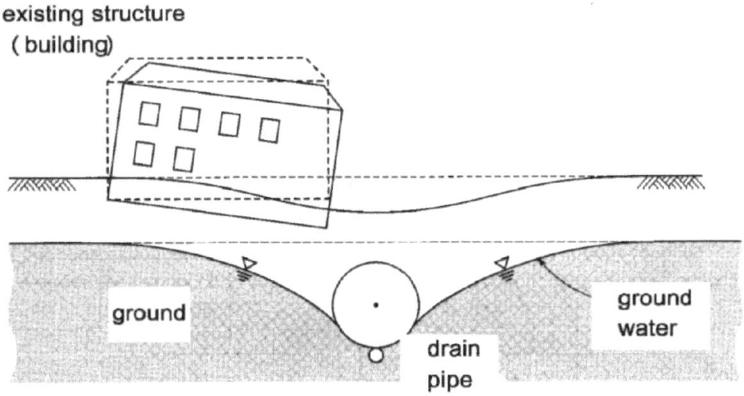

Figure 28. Effect of the tunnel acting as a drain (Cost C7 2003).

The settlements predicted in the design were about 110 mm, even though the differential settlements were not very significant. This fact was justified by the existence of layers of calcareous sandstone with about 4 m thickness. Nevertheless, after the excavation of a first cavity, it has been verified that the differential settlements were much higher than those predicted by the numerical models, which has caused an increase in the cracking in old buildings existing at the surface. As a result, compensation grouting has been performed during the execution of a second cavity. This has made it possible to limit the settlements at the surface to the values predicted in the design and to reduce the damages in the buildings.

The station consists of two large parallel caverns with about 18 m span and 250 m length with a cross-section of 240 m^2, separated by a minimum distance of about 6 m. The NATM method was used for its construction. The west station was crossed by the shield driven machine, due to construction reasons related with planning of the works (Figure 30). Figure 31 presents a geological profile of the west station.

(a)

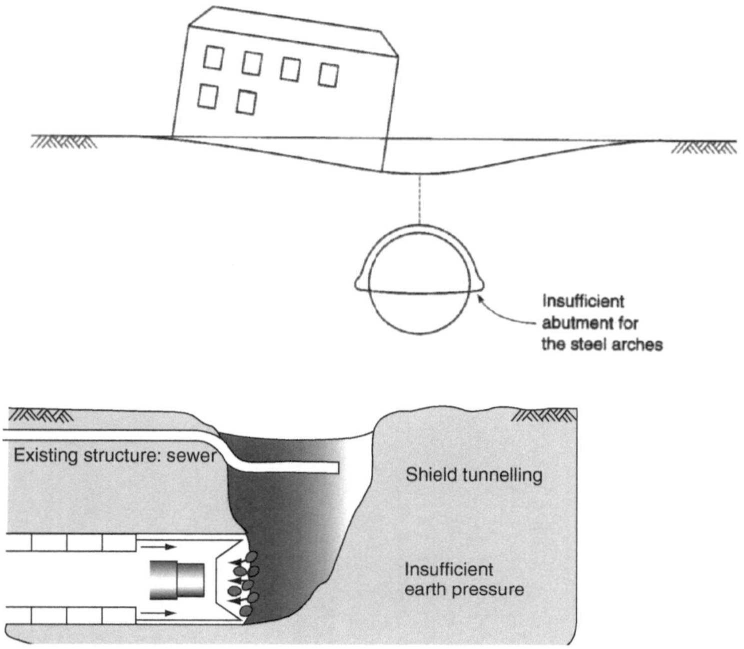

Figure 29. Tunnel face instability associated with lack of integrity of the tunnel lining (Cost C7 2003).

The selection of the construction sequence has been determined by the experience of other works and by geotechnical research carried out on basis of numerical models using the software FLAC. Very complex calculation meshes were used with a high number of calculation stages, as well as the use of the convergence-confinement method was adopted. Several alternative structural solutions have been analyzed, in order to permit a more appropriate decision making as regards the analysis of the stability of the successive underground structures. Figure 32 illustrates the final excavation sequence adopted.

A basic monitoring programme of the station and adjacent buildings was elaborated to assess the stability of the different construction phases and to detect anomalous behaviours in the surrounding buildings, especially to detect the damages occurred. The aim of the instrumentation plan was to control the geotechnical works and the buildings at the surface through:

i) measurement of displacements inside the mass in the vicinity of excavations by means of rod-extensometers and inclinometers;
ii) measurement of superficial settlements and superficial displacements;
iii) measurement of convergence on the supports;
iv) reading of the variation of hydrostatic pressures by means of piezometers;
v) measurement of crack movements in installations at the surface.

Eleven main observation sections have been defined for the station with about 25 m spacing, The convergences has been measured on basis of readings in three marks according to triangles, the position of the lower marks varying in accordance with the construction stage. The reading frequency varied according to the distance of the excavation head of the instrumented section. Measurements of readings of piezometric loads in piezometers have been done. As regards the surface, levelling measurements have been performed on topographic marks per each section. Other marks have also been used for topographic levelling at the surface of the soil. Topographic reading gauges placed on the walls of the buildings and topographic reading targets placed also on buildings have also been used. For the control of cracking, seals have been placed on buildings

Figure 30. Baixa-Chiado station.

in order to facilitate the reading of the opening of cracks whenever they occurred in the buildings due to the excavation of cavities.

Figure 33 shows settlements monitored for the cross-section P33 on two instants, 1995/08/28 and 1997.04.30, which are compared with numerical simulations. Those values are compared with values obtained by the numerical simulations. At the first referred instant the shield had already crossed the west cavity and it was initiated the bench of the east cavity. At the second instant both cavities had already been excavated. Due to the influence of the compensation grouting the observed values in the surface did not change significantly.

At the surface station area, there are about 30 old buildings between the *Largo da Academia das Belas Artes* and the *Calçada do Livramento*. For assessing the characteristics of the buildings, inspection visits have taken place before the beginning of construction. It has been verified that the construction adopted by the Portuguese authorities, of the prototype *Pombalina*, is the most generalized type of construction. It was implemented immediately after the Lisbon earthquake in 1755, which was responsible for the destruction of a large area of the downtown part of Lisbon. An association of resistant masonry elements with a structure with the configuration of a timber cage characterizes that construction solution. The resistant masonry forms the peripheral walls as

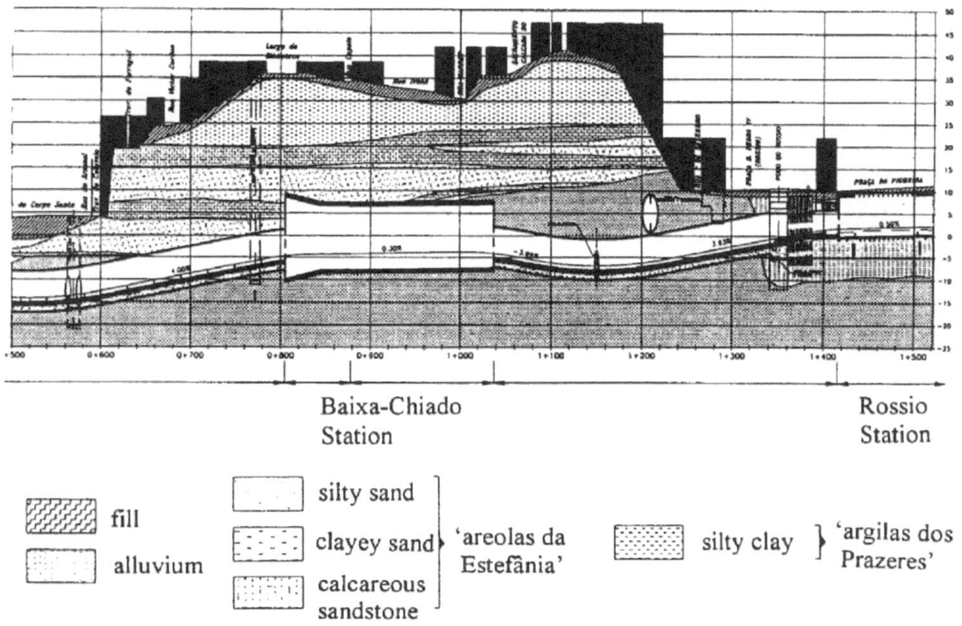

Figure 31. Geological profile of *Baixa-Chiado* station.

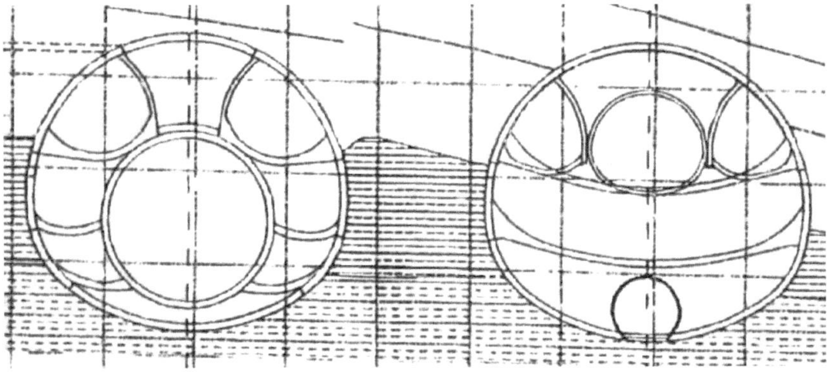

Figure 32. Excavation sequence adopted for the construction of caverns.

well as some walls of the lower floors and it is supported on continuous foundations. The timber structure is continuous in every plan and comprises the resistant elements of the pavements and internal division walls. It is connected continuously to timber parts embedded in the surrounding walls. It has been nevertheless verified that numerous alterations have been performed throughout the years, especially in lower floors.

There are also other prototypes of construction, such as palaces and churches, without cage effects. The occurrence of some buildings with reinforced concrete structures has also been verified, particularly in buildings recently rehabilitated after the occurrence of a great fire at *Chiado*. Some of the buildings presented initially major anomalies related with the alterations introduced during their lifetime, namely resulting from cuts performed in the cage structures, for the creation of large commercial spaces and the execution of caves. Therefore, the condition of

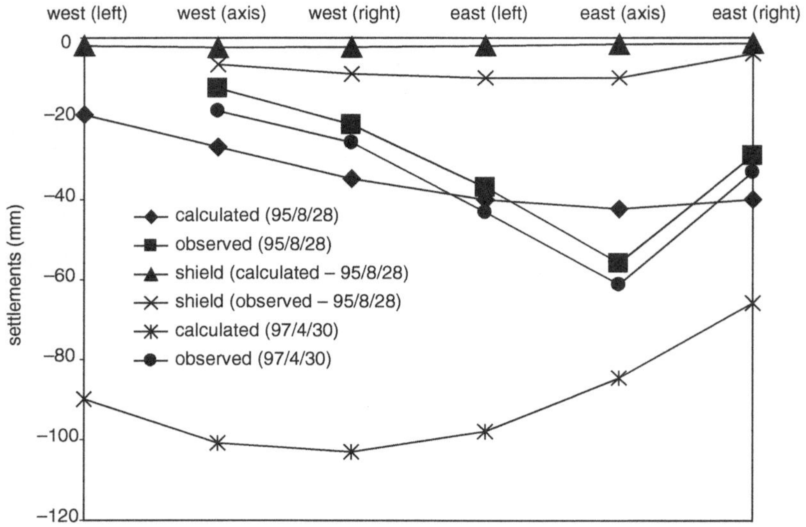

Figure 33. Settlements for section P33.

those buildings had become vulnerable as regards horizontal and vertical actions and differential settlements of the foundations. Thus, hazard levels were established for the various buildings involved. A group of 6 buildings submitted to high hazard was considered for the purpose. Those buildings were subsequently the object of a careful observation and some of them were submitted to consolidation works.

Empirical laws have been used to estimate the settlements on basis of the monitoring results, considering Gaussian functions characterized by the maximum settlement on the center line S_{max} and the horizontal distance from the cavity centre-line to the point of inflection of the settlements i. The following parameters have also been considered: relative volume of surface settlement $\Delta V/V$, which depends on the geotechnical conditions and on the construction method, and $K = i/z$ that depends mainly on the geotechnical characteristics of the ground, z being the depth of the axis of the tunnel (Barreto et al. 1999). Predictions have been performed for several instrumentation profiles adopting relevant construction stages. Using the approach of Gaussian curves, large discrepancies can be verified in same cases. If the selected type of curve does not fit the data points well, significant errors may affect ground distortions evaluation. In these cases curves of yield density type can adjust better to settlement data points. On basis of the parameters calculated, the potential damages in the buildings for the final construction stage have been estimated.

For the parameters assessment using the Gaussean curves, the most relevant construction stages have been chosen, namely considering:

i) east station – execution of the head, bench and invert to complete section;
ii) west station – excavation with shield. The variation of the parameters $\Delta V/V$ and K was analysed according to the construction stages and to the characteristics of the geotechnical profile.

A zoning of the area enveloped by the station has been done according to the most significant empirical parameters obtained. The following conclusions have thus been obtained:

i) The relative volume loss of ground corresponding to the excavation with shield is usually less than the cavities according to the NATM with few exceptions.
ii) The relative volume loss of ground in the excavations NATM decrease in the bench of the section, from 1.25 to 0.9% for the first monitored section, and from 0.7 to 0.45% for another section.

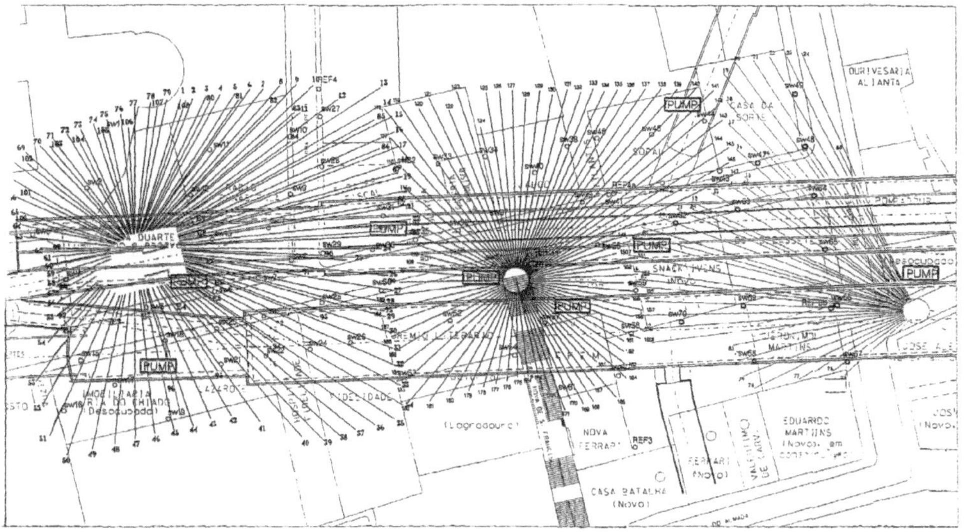

Figure 34. Compensation grouting at Baixa-Chiado station.

iii) The shield excavation shield leads to additional losses in the NATM cavity. In profile P33 a 0.85% loss in the first stage is produced, which is increased to 1.1% after the passage of the shield in the other cavity.
iv) The magnitude of the relative volume loss of ground is highly influenced by the type of ground, for sandy formations losses of 0.8–1.25% can be found whereas for clayish formations coefficients of 0.4–0.7% can be observed.
v) The horizontal distance from the tunnel centre-line to the point of inflexion of the settlement presents usually sandy formations characteristic values, i.e., $0.3 < K < 0.4$. For the assessment of damage, values of relative volume loss of ground have been adopted for the various sections, ranging from 0.6 to 1%, K varying from 0.3 to 0.4.

Based on the assessed values, the potential damage in the buildings has been calculated considering the damage criteria suggested by Boscardin and Cording (1989) and by AFTES (1995). The maximum intensity of damage has been considered moderate, i.e., type 3, in accordance with the Boscardin and Cording classification. After the excavation of the first station, cracks have occurred in some buildings. Reinforcement measures of some of these structures have been taken and other measures have also been adopted.

The maximum settlement resulting from the excavation of the east station has been very close to the predictions done. Nevertheless, the same did not occur with the angular distortions that were higher than what had been expected, approximately 1/300. The predictions for the station indicated, therefore, an increase of the angular distortions and the occurrence of settlements additional to those due to the excavation of the east station. This fact was likely to produce significant damage in the old buildings. Thus, a process of compensation grouting has been implemented, which has made it possible to create a structure similar to a slab that compensated partly the deformations induced by the excavations of the cavities (Figure 34).

6 CONCLUSIONS

The considerations presented above lead to conclude that the tunnels are examples of geotechnical works that may cause accidental situations with serious consequences for workers, for the infrastructures in the vicinity and for the environment.

The most severe and relatively frequent situations of accidents in tunnels are related with the collapse of the excavation front, with the destruction and deformation of supports, with flooding, fire, explosions and other emergency situations. The interaction with the infrastructures on the surface and on the subsurface is an aspect to be taken into account in the analysis of the risk and safety of these works.

Several recommendations can be established, with a view to minimize and prevent accidental events during construction, operation and rehabilitation of tunnels, so as to improve safety and durability of these underground structures and to reduce costs, with the help of innovative technologies. The risk analyses in the design of tunnels are essential in all stages of the design, construction and operation.

REFERENCES

AFTES 1995. Settlements due to the excavation of underground work. Recommendations of the Working Group no. 8: 373–395.

Babendererde, L. 1999. TBM drives in soft ground – Weak points in process engineering. ITA Congress on *Challenges for the 21st Century*, Oslo: 811–815.

Barradas, J.; Sousa, L.R.; Horta, S. 2002. Structural analysis and the observed behaviour of the Grilo tunnel and adjacent slopes (in Portuguese). 8th Portuguese Geotechnical Congress, Lisbon: 1385–1398.

Barreto, J.; Fernandes, D.; Sousa, L.R.; Cardoso, A.S. 1999. Field Observation of the Baixa-Chiado Station, Lisbon Metro. ITA Congress on *Challenges for the 21st Century*, Oslo: 3–12.

Barton, N. 2000. TBM tunneling in jointed and faulted rock. Ed. Balkema, Roterdam: 172.

Boscardin, M.D.; Cording, E.G. 1989. Building response to excavation-induced response. J. Geotechnical Engineering, ASCE, Vol. 115, no. 1: 1–21.

Brekke, T.L.; Ripley, B.D. 1987. Design guidelines for pressure tunnels and shafts. Electric Power Reserch Institute, Report EPRI AP-5273, Berkeley.

Burland, J.B.; Standing, J.R.; Jardine, F.M. 2002. Assessment the risk of building damage due to tunneling – lessons from the Jubilee Line Extension, London. 2nd Conference on *Soil Structure Interaction in Urban Civil Engineering*, Zurich: 11–38.

Caldeira, L. 2002. The risk analysis and uncertainties (in Portuguese). 8th Portuguese Geotechnical Congress, Lisbon: 2295–2317.

Cost C7 2003. Avoiding damage caused by soil-structure interaction: lessons learnt from case histories. Thomas Telford, Ed. Kastner et al., London: 77.

Dudt, J.P.; Descoeudres, F.; Einstein, H.; Egger, P. 2000. Decision Aids for Tunnelling applied to enterprise biding comparison (in French). 2nd Int. Conference on *Decision Making in Urban and Civil Engineering*, Lyon.

Einstein, H. 2002. Risk assessment and management in geotechnical engineering. 8th Portuguese Geotechnical Congress, Lisbon: 2237–2262.

Einstein, H.; Intermitte, C.; Sinfield, J.; Descoeudres, F.; Dudt, J. 1999. Decision aids for tunneling. Transportation Research Rep. no. 1656, National Academy Press, Boston.

Freitas, V.; Malva, R.; Sousa, L.R.; Oliveira, M. 2003. Portuguese experience in inspection of railway tunnels (in Portuguese). Portuguese-Spanish Symposium on *Underground Works – Relevance of Survey and Geotechnical Observation*, Madrid: 347–358.

Grasso, P.; Xu, S.; Fedele, M.; Russo, G.; Chiriotti, E. 2003. Particular failure mechanisms of weathered granite observed during construction of metro tunnels by TBM. ITA Congress on *(Re)Claiming the Underground Space*, Amsterdam: 497–503.

Gysel, M. 2002. Anhydrite dissolution phenomena: Three case histories of anhydrite karst caused by water tunnel operation. Rock Mechanics and Rock Engineering, Vol. 35, no. 1: 1–21.

Hansen, L. 1993. The significance of general and structural geology in Rock Engineering. ISRM Symposium EUROCK '93 on *Safety and Environment Issues in Rock Engineering*, Lisbon: 811–816.

Hoek, E. 2000. Rock engineering. Course notes, Rocscience, www.rocscience.com.

Hoek, E.; Bray, J. 1973. Rock slope engineering. Institution of Mining and Metallurgy, London: 358.

HSE 1996. Safety of New Austrian Tunnelling Method (NATM) tunnels. Health & Safety Executive, London: 86.

ICE 1996. Sprayed concrete linings (NATM) for tunnels in soft ground. Institution of Civil Engineers, Report, London: 88.

Inokuma, A. et al. 1994. Studies on present state and the mechanism of trouble occurrence in tunnel construction in Japan. ITA Conference, Cairo: 239–246.
Lamas, L.N. 1993. Contributions to understanding the hydromechanical behaviour of pressure tunnels. PhD Thesis, Imperial College, London: 419.
Lemos, J.V.; Sousa, L.R.; Ramos, J.M.; Barradas, J. 2003. Static and dynamic structural analysis of Grilo tunnel, Portugal. 4th International Workshop on *Applications of Computational Mechnics in Geotechnical Engineering*, Ouro Preto: 75–83.
Martins, J.B.; Sousa, L.R.; Barreto, J.; Martins, F. 2003. Excavation of transport urban tunnels with TBM EPB in heterogeneous granite formations and in pre-consolidated soil formations (in Portuguese). Portuguese-Spanish Symposium on *Underground Works – Relevance of Survey and Geotechnical Monitoring*, Madrid: 449–458.
Matos, A.C. 1999. The geometric heterogeneous of rock masses and the underground works. The case of the Porto planned undertakings (in Portuguese). Journal *Ingenium*, July: 16–18.
Neto, F.; Kochen, R. 2002. Safety, rupture and collapse of urban NATM tunnels (in Portuguese). 4th Symposium on *Tunnels in Urban Environment*, São Paulo: 47–52.
Ortega, R. 2003. Stabilization by anchors of the cut-and-cover Urritza tunnel (Navarra), (in Spanish). Portuguese-Spanish Symposium on *Underground Works – Relevance of Survey and Geotechnical Monitoring*, Madrid: 317–321.
Rocha, M. 1977. Some problems related to the Rock Mechanics of low strength materials (in Portuguese) Journal *Geotecnia*, no. 18: 3–27.
Silva, C.O. 2001. Safety control of railway tunnels. Development of methodologies based in KBS systems (in Portuguese). MSc thesis, University of Porto: 276.
Silva, H.S. 1988. Hydromechanical behaviour of concrete dam foundations. Application to Cahora-Bassa dam. MSc thesis, Lisbon: 185.
Silvestre, M.; Sousa, L.R.; Hack, R. 2002. Laboratory study of geomectrical and hydromechanical characteristics of discontinuities. ISRM News Journal, Vol. 7, no. 2: 9–15.
Sousa, J.A. 1998. Tunnels in soil formations. Behaviour and modeling (in Portuguese). PhD Thesis, University of Coimbra: 623.
Sousa, L.R. 2000. Innovative aspects in the design and construction of underground structures (in Portuguese). 7th Portuguese Geotechnical Congress, Porto: 1313–1373.
Sousa, L.R. 2001. Observation in the safety control of underground structures in urban environment (in Portuguese). SPG, Course on *Tunnels in Urban Area*, Coimbra: 56.
Sousa, L.R.; Lamas, L.N.; Martins, C.S. 1994. Applications of computational mechanics to underground structures in hydraulic projects. 1st Worskshop on *Applications of Computational Mechanics in Geotechnical Engineering*, Rio de Janeiro: 15–88.
Sousa, L.R.; Ramos, J.M.; Silva, H.S. 1995. Cahora-Bassa hydroelectric scheme: A new monitoring plan. 8th ISRM Congress, Tokyo: 861–864.
Sousa, R.L.; Einstein, H.; Correia, A. 2004. Assessment and improvement of risk analysis methodologies for tunneling projects. MIT, Plan of PhD Thesis, Cambridge.
Uenishi, K.; Sakurai, S. 2002. Earthquake-induced collapse of the Daikai underground station in Kobe: evaluation of the characteristics of the associated seismic waves. Symposium NARMS-TAC 2002 on *Mining and Tunneling Innovation and Opportunity*, Toronto: 767–774.
Vlasov, S.N.; Makovsky, L.V.; Merkin, V.E. 2001. Accidents in transportation and subway tunnels. Construction to operation. Russian Tunnelling Association, Moscow: 198.

Geotechnical Risks in Rock Tunnels – Campos e Matos, Ribeiro e Sousa, Kleberger & Lopes Pinto (eds) © 2006 Taylor & Francis Group, London, ISBN 0 415 40005 8

Geological risk in the use of TBMs in heterogeneous rock masses – The case of "Metro do Porto"

S. Babendererde
Babendererde Ingenieure GmbH, Germany

E. Hoek
Consultant, Canada

P. Marinos
National Tecnical University of Athens, Greece

A. Silva Cardoso
University of Porto, Portugal

ABSTRACT: The highly variable nature of the deeply weathered Oporto granite posed significant challenges in the driving of the 2.3-km-long C line and the 4-km-long S line of the Oporto Metro project. Two 8.7-m diameter Herrenknecht EPB TBMs were used to excavate these tunnels and early problems were encountered due to over excavation and face collapse. Three collapses to surface occurred in the first 400 m of the C line drive and the last of these collapses resulted in the death of an occupant of a house on the surface. This resulted in a delay of almost 9 months to the project and the difficulty was finally resolved by the introduction of an Active Support System which involves the injection of pressurized bentonite slurry to compensate for deficiencies in the face support pressure when driving in mixed face conditions. Both the C and S lines have now been completed with minimal surface subsidence and no face instability.

1 INTRODUCTION

In late 1998 the Municipality of Porto took a decision to upgrade its existing railway network to an integrated metropolitan transport system with 70 km of track and 66 stations. Seven kilometers of this track and 10 stations are located under the picturesque and densely populated city of Porto, a UNESCO world heritage site. A map of the surface and underground routes is presented in Figure 1.

Metro do Porto SA, a public company, is implementing the project with a concession period of 50 years. The design, construction, and operation of this concession was awarded to Normetro, a joint venture including civil contractors, equipment and system manufacturers, and an operator. The civil works design and construction was awarded to Transmetro, a joint venture of Soares da Costa, Somague and Impregilio.

The underground tunnel, driven by two earth pressure balance (EPB) TBMs, has an internal diameter of 7.8 m and accommodates two tracks with trains running in opposite directions. Line C stretches 2350 m from Campanhã to Trindade and has five underground stations, a maximum cover of 32 m and a minimum of 3 m before reaching Trindade station. Line S is 3950 m long and runs from Salgueiros to São Bento with 7 stations and a maximum overburden of 21 m.

Tunnel driving was started in August 2000 with the drive from Campanhã to Trindade. It was originally planned that the EPB TBM would be run with a partially full, unpressurized working chamber in the better quality granite in order to take advantage of the higher rates of advance in this mode as compared with operating with a fully pressurized working chamber. It was soon found that the highly variable nature of the rock mass made it extremely difficult to differentiate

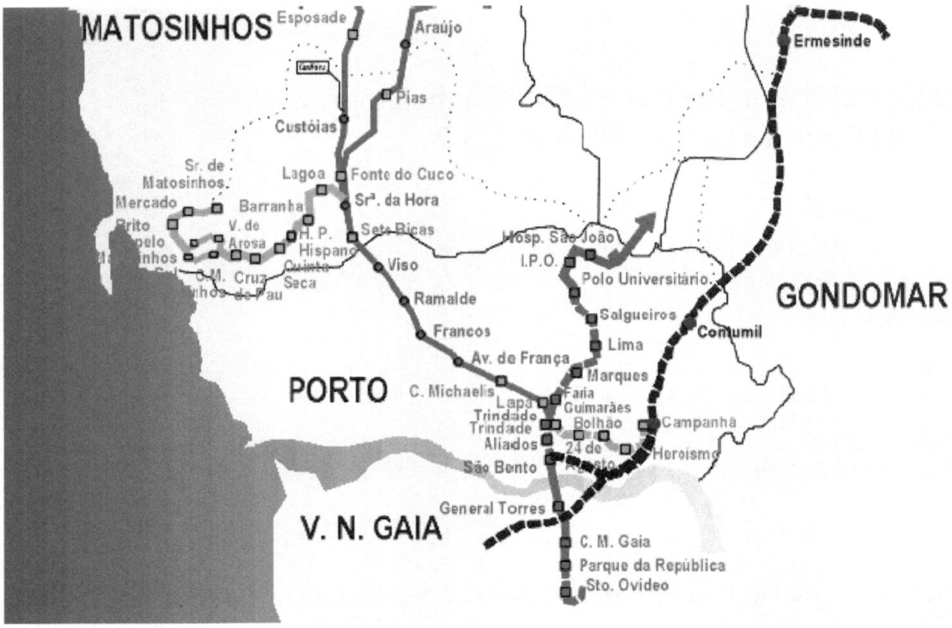

Figure 1. Map of Metro do Porto routes. Underground tunnels are line C from Campanhã to Trindade and line S from Salgueiros to São Bento.

between the better quality rock masses in which the working chamber could be operated safely with no pressure and the weathered material in which a positive support pressure was required on the face. There were indications of overexcavation and there were a two collapses to surface. The first occurred during excavation on 22 December 2000 between segments no. 318 and 327. The second was located between segments 291 and 298 and occurred on 12 January 2001, almost a month after the passage of the TBM on 16–18 December 2000. This collapse resulted in the death of a citizen in a house overlying the tunnel.

At the invitation of Professor Manuel de Oliveira Marques, Chief Executive Officer of Metro do Porto S.A., one of the authors (E.H.) visited Porto from in early February 2001 to review the geotechnical and tunnelling issues of the C line tunnel. As a result of this visit a Panel of Experts, consisting of the authors of this paper, was established and met in June 2001 in order to provide advice on the restart of the TBM drive.

2 GEOLOGICAL CONDITIONS

The underground portion of the line passes through the granite batholith which was intruded into the Porto–Tomar regional fault in the late Hercinian period (Figure 2). The Porto granite, a medium grained two mica granite, is characterized by deep weathering and the tunnel passes unevenly through six grades or weathering and alteration ranging from fresh granite (W1) to residual soil (W6). The granite is crossed randomly by aplitic/pegmatitic dykes which display much less weathering, following tectonically determined tension joints.

The particular feature of most engineering significance of the rock mass is its weathering. All weathering grades (W1–W6, as established in the engineering geological classification according to the scheme proposed by the Geological Society of London, 1995) can be found. The depth of weathering is of the order of few tens of meters as weathering was assisted by the stress relief regime due to the deepening of the Douro valley. Depths of weathering of 30 m are reported in Begonha and Sequeira Braga, 2002. Hence, the ground behaviour varies from a strong rock mass

Figure 2. Distribution of granite in the city of Oporto (from A. Begonha and M. A. Sequeria Braga, 2002).

Figure 3. Appearance of different degrees of weathering in granite in a core recovered from a site investigation borehole on the tunnel alignment. Note that the weathered granite in the left box is at a depth of about 24 m under the sound granite of the right box. This must therefore correspond to a huge boulder (core).

to a low cohesion or even cohesionless granular soil. The granularity and frictional behaviour is retained as the kaolinitisation of feldspaths is not complete and the clay part not important. Furthermore, the spatial development of the weathered rock is completely irregular and erratic. The change from one weathered zone to another is neither progressive nor transitional. It is thus possible to move abruptly from a good granitic mass to a very weathered soil like mass. The thickness of the weathered parts varies very quickly from several meters to zero. Weathered material, either transported or *in situ*, also occurs in discontinuities.

A particularly striking feature is that, due to the erratic weathering of the granite, weathered zones of considerable size could be found under zones of sound granite (see Figures 3 and 4).

While this phenomenon is an exception rather than the rule and it was expected to disappear with depth, it could not be ignored in the zone intersected by the construction of the metro works. A typical case of such setting is in Heroismo station where weathered granite with floating cores of granite occurs under a surficial part of a sound granitic rock mass (Figure 5).

The permeability of the rock mass is dependent upon the weathering grade. In the less weathered rock the flow is related primarily to the fracture system while, in the more heavily weathered material, the ground behaves more like a porous medium. Porosity in the latter case may have been increased from leaching. Water supply, in the past, was by means of a large number of wells within the city and these, together with the highly variable permeability of the rock mass, have resulted in a very complex ground-water regime. The overall permeability is rather low of the order of 10–6 m/s or lower. However good permeabilities were measured in pumping tests. We consider that preferential drainage paths exist within the granite mass. The very weathered material having little or no cohesion may be erodable under high hydraulic gradients.

Figure 4. Appearance of Oporto granite in the face of an excavation for the new football stadium. Fracturing of the rock mass and heterogeneity of weathering is obvious.

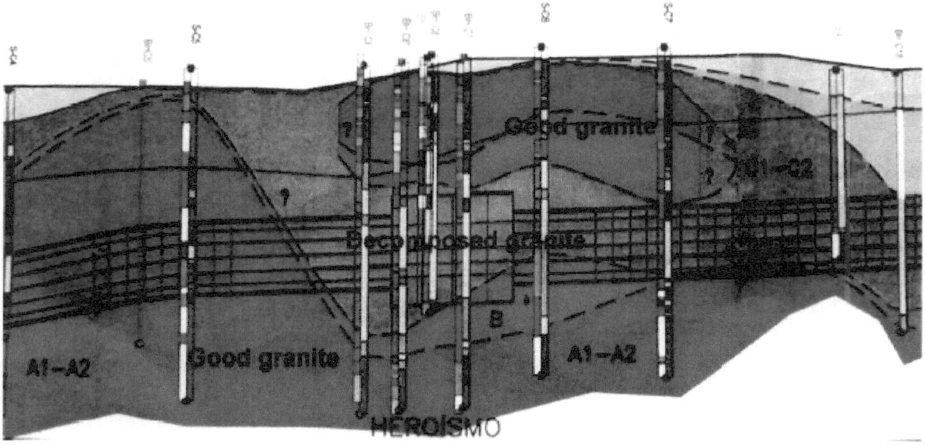

Figure 5. Predicted geology for the Heroismo mined station (assessment by Transmetro, documents of Metro do Porto). Hetergeneity in weathering and its erratic geomentry is evident.

3 EPB TBM CHARACTERISTICS

The complex geological and hydrogeological conditions described above resulted in a decision by Transmetro to utilize an 8.7-m-diameter Herrenknecht EPB TBM (see Fruguglietti et al. 1999, 2001). Initially, only one machine was to be used to drive both lines but following start-up problems, a second machine was added in order to make it possible to complete the tunnel drives on schedule.

The TBMs are equipped with a soil conditioning system capable of injecting foam, polymer or bentonite slurry into the working chamber. Muck removal is by continuous belt conveyor from the TBM back-up to the portal and then by truck to the muck disposal areas. Tunnel lining is formed from 30-cm thick, 1.4-m wide precast concrete segments. The lining comprises six segments and a key and dowel connectors are used in the radial joints while guidance rods are used in the longitudinal joints. The features of the EPB TBM are illustrated in Figure 6. In a review paper by N. Della Valle (Tunnels and Tunnelling, 2002) details are presented. In that paper issues proposed by the authors of the present paper are also described.

4 IMPLICATIONS OF GEOLOGICAL CONDITIONS IN TERMS OF THE TBM OPERATION

The geological conditions discussed above can be translated to the following geological models in front, at the face (Figure 7) and immediately above the TBM:

1. Granitic mass of sound or slightly weathered rock, no weathered material in the discontinuities;
2. Granitic mass of sound or slightly weathered rock but with very weathered material (filled or *in situ*) in substantial fractures; these fractures may communicate with overlaying parts of completely weathered granite;

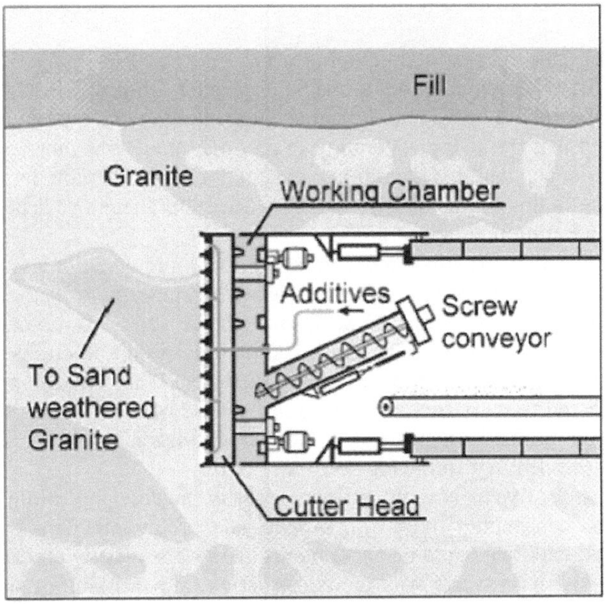

Figure 6. Characteristics of the Herrenknecht EPB TBM used in Oporto.

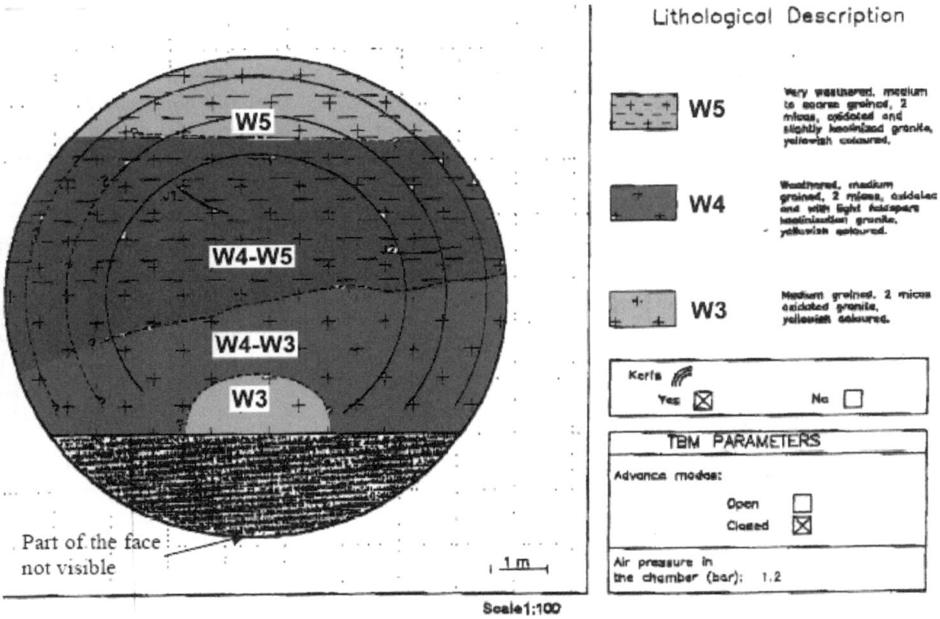

Figure 7. As typical distribution of weathered granite in the face of the EPB driven tunnel.

3. Very weathered or completely weathered granite, W5 (almost granular soil with little or no cohesion);
4. Very weathered or completely weathered granite with blocks of the rock core;
5. Mixed conditions with both sound mass and completely weathered granite appearing in the face.

In all cases the water table is above the tunnel crown.

Only the first of these geological models can be excavated using an EPB TBM operating in an open mode. However, because of the unpredictable changes in the geological conditions described above, we considered that the risk of operating in an open mode was unacceptable unless there was unambiguous evidence that this condition persisted for a considerable length of tunnel drive. This was not the case in this tunnel and we recommended that the entire drive should be carried out with the TBM operating in a closed mode.

Indeed in all other models, uncontrolled overexcavation could occur unless the chamber of the machine was full of appropriately conditioned excavated material with the necessary pressurisation and control of the evacuation of the muck through the screw conveyor. Lack of adequate face support could result in piping of the weathered material in the fractures that could, in turn, induce collapse of the overlying weathered granite. The mixed face conditions described in item 5 above were considered to be particularly difficult because of the uneven pressure distribution on the face induced by the different stiffness of the rock and soil masses. The successful handling of this problem is discussed in a following section.

A significant number of wells and old galleries exist in the area and, while most were located on old city maps and by inspection of existing properties, there remained the possibility that some unpredicted wells and galleries could be encountered. The wells usually end above the tunnel but some were deep enough to interfere with the construction. The crossing of such features clearly involved some risk but this was substantially lower when operating the TBM in a fully closed and pressurised mode than in an open or partially open mode.

5 FACE SUPPORT PRESSURE

The face support pressure of EPB – TBMs is controlled by measuring the pressure at the bulkhead with pressure cells, approximately 1.5 m from the face, as shown in Figure 8. In closed mode operation, the working chamber is completely filled with conditioned excavated material, the earth paste. The earth paste is pressurized by the advancing forces induced by the advance jacks via the bulkhead. The pressure level is controlled by the effectiveness of the excavating cutter head in relation to the discharging screw conveyor.

To verify complete filling of the working chamber, the density of the earth paste in the working chamber is controlled by pressure cells on the bulkhead at different levels. This method satisfies the demand of preventing a sudden instability of the face caused by a partially empty working chamber but it does not guarantee a reliable face support pressure.

Pressure measurement at the bulkhead, 1.5 m behind the face, provides only partial information about the support pressure at the face. The support medium, the earth paste created from excavated ground, conditioned by a suspension with different additives, must have the physical properties of a viscous liquid. However, the shear resistance in that viscous liquid reduces the support forces which can be transferred onto the face. The shear resistance of the earth paste depends on the excavated ground and the conditioning, which is a complex and sensitive procedure. Consequently, the shear resistance of the support medium often varies considerably.

Therefore, the fluctuation of the face support pressure can exceed 0.5 bars. This fluctuation may be acceptable in homogeneous geology but in mixed ground, as found in the Oporto granite, the variable support pressure entails the danger of significant over excavation.

One of the processes which can cause a drop in the face support pressure is illustrated in Figure 9 which shows a situation in which the lower part of the face is in unweathered granite while the upper part of the face is in residual soil. A major part of the thrust of the machine is consumed by the cutter forces required to excavate the unweathered granite and there is a deficiency in the forces available to generate the pressure in the earth paste in the upper part of the working chamber. This results in an imbalance between the soil and water pressure in the unweathered granite and the support pressure in the upper part of the working chamber. If this deficiency is too large, the face will collapse inwards into the working chamber and this will result in progressive overexcavation ahead and above the face.

The deficiency of face support pressure can be compensated for by the addition of an Active Support System, proposed by Dr Siegmund Babendererde (one of the authors of this paper) and shown in Figure 9. This system is positioned on the back-up train and consists of a container filled with pressurized bentonite slurry linked to a regulated compressed air reservoir. The Bentonite slurry container is connected with the crown area of the working chamber of the EPB TBM. If the

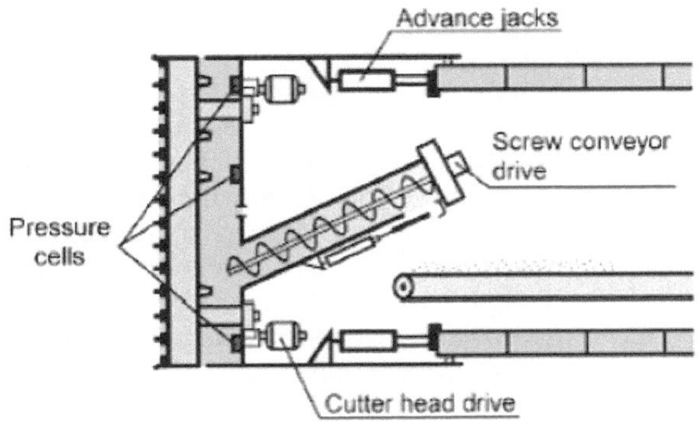

Figure 8. Measurement devices for face support pressure.

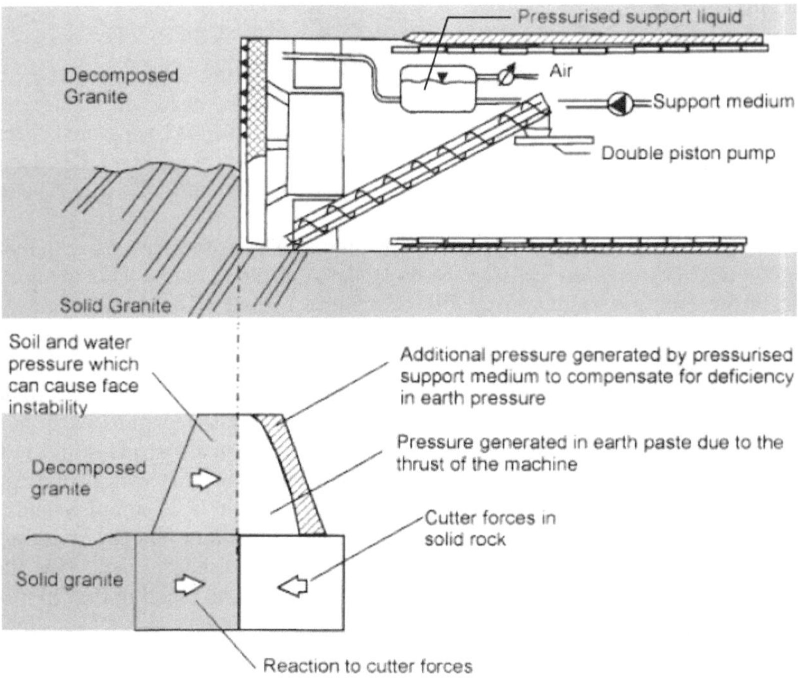

Figure 9. Face support pressures in mixed face conditions in Oporto granite. An Active Support System for overcoming the support pressure deficiency is also illustrated.

support pressure in the working chamber drops below a predetermined level, the Active Support System automatically injects pressurized slurry until the pressure level loss in the working chamber is compensated. The addition of this Active Support System to the EPB TMB results in an operation similar to that of a Slurry TBM. This automatic pressure control system reduces the range of fluctuations of the face support pressure to about 0.2 bar.

In the case of an open and potentially collapsible structure in the weathered granite surrounding the wells, resulting from leaching of the fines, we considered that stable face conditions could be maintained by the correct operation of the TBM in fully closed EPB mode with supplementary fluid pressure application. However, care was required in the formulation and preparation of the pressurizing fluid in order to ensure that an impermeable filter cake was formed at the face. This was necessary in order to prevent fluid loss into the open structure of the leached granite mass.

The application of the Active Support System in the Metro do Porto project was the first time that this system had been used. There was initial concern that the addition of the bentonite slurry would alter the characteristics of the muck to the point where it could no longer be contained on the conveyor system and that an additional slurry muck handling facility may be required. This concern proved to be unfounded since the volume of bentonite slurry injected proved to be very small and there was no discernable change on the characteristics of the muck.

The predetermined support pressure was determined from calculations using the method published by Anagnostou and Kovari (1996) which proved to be reliable for these conditions. The Active Support System was extremely effective in maintaining the predetermined support pressure and no serious face instability or overexcavation problems were encountered after it was introduced. In fact, the system permitted the 8.7-m-diameter tunnel to pass under old houses with a cover of 3 m to the foundations, without any pretreatment of the ground. Surface settlements of less than 5 mm were measured in this case.

The Active Support System was also connected to the steering gap around the shield and the filling of this gap with bentonite slurry provided a reliable means of maintaining a predetermined pressure in this gap.

6 TBM CUTTER WEAR

The Oporto granite is a highly abrasive material and, when broken down into the conditioned paste in the working chamber of an EPB TBM, it becomes a very effective grinding paste. Even in the very weathered form, the final weathering product contains the skeleton minerals, mainly quarts, which characterise the dominant granular fraction. For example, the average percentage of secondary minerals in both the bulk weathered rock or the bulk saprolite is low ~2.5%, and ~9%, respectively (A. Begonha and M.A. Sequeira Braga, 2002). This keeps abrasiveness high and proved to be a major problem in Oporto where cutter wear necessitated frequent replacement of cutters. On average, one cutter was consumed per running metre of tunnel drive and, in addition, there was severe wear of the cutter head construction. These problems are illustrated in Figures 10 and 11.

Numerous trials were carried out with different conditioning agents in an attempt to reduce cutter wear but none of these proved to be of any great help and the cutter wear problem persisted until the end of the tunnel drives. Some relief from the bearings and bearing mounts was obtained by welding a deflector wedge ahead of the cutter assembly as shown in Figure 12.

7 CONCLUSIONS

The highly variable characteristics of the weathered granite in Oporto and their sudden changes imposed substantial risks on the driving of the C and S lines by means of EPB TBMs. The impossibility of accurately predicting and maintaining the correct face support pressure resulted in significant overexcavation and two collapses to surface during the first 400 m of the C line drive.

The introduction of the Active Support System, which involves the injection of pressurized bentonite slurry to compensate for deficiencies in the face support pressure when driving in mixed face conditions, proved to be a very effective solution. The remaining C and S line drives have now been completed without further difficulty although the rate of progress was less than that originally projected when the project was planned.

The final breakthrough of the C line drive is illustrated in Figure 13.

Figure 10. Typical wear of the disk cutter showing that the flanks of the disk are worn as quickly as the cutting surface.

Figure 11. Wear of the disk cutter bearings and bearing mounts. Disks that do not rotate freely wear asymmetrically while freely rotating disks are abraded as shown in Figure 10.

Figure 12. Addition of deflector wedges ahead of disk cutters helped to deflect paste from the bearing assembly.

Figure 13. Final breakthrough of the TBM S-203 on the completion of the drive from Salgueiros to Trindade on Thursday 16 October 2003.

ACKNOWLEDGEMENTS

The authors wish to acknowledge the permission of Metro do Porto to publish the details contained in this paper. The cooperation of Transmetro and particularly of Ing. Giovanni Giacomin in working with the Panel of Experts is also acknowledged.

REFERENCES

Anagnostou, G. and Kovari, K. 1996. "Face stability conditions with earth-pressure balanced shields". Tunnelling and Un-derground Space Technology. Vol 11, No. 2, pp. 163–173.

Begonha, A. and Sequeira Braga, M. A. 2002. "Weathering of the Oporto granite: geotechnical and physical properties". Catena. Vol. 49, pp. 57–76.

Della Valle, N. 2002. "Challenging soil conditions at Oporto". Tunnels and Tunnelling International. December 2002, pp. 16–19.

Fruguglietti, A., Guglielmetti, V., Grasso, P., Carrieri, G. and Xu, S. 1999. "Selection of the right TBM to excavate weathered rocks and soils". Proceedings Conference: Challenges for the 21st Century, Allen et al. (eds), Balkema Publ., pp. 839–947.

Fruguglietti, A., Ferrara, G., Gasparini, M. and Centis, S. 2001. "Influence of geotechnical conditions on the excavation methods of Metro do Porto project". Proceedings Congress ITA, Milan, pp. 135–141.

Geological Society of London 1995. "The description and classification of weathered rocks for engineering pur-poses". QJEG, p. 28.

Russo, G., Kalamaras, G.S., Origlia, P. and Grasso, P. 2001. "A probabilistic approach for characterizing the complex geological environment for the design of the new Metro do Porto". Proceedings Congress ITA, Milan, pp. 463–470.

Geotechnical Risks in Rock Tunnels – Campos e Matos, Ribeiro e Sousa, Kleberger & Lopes Pinto (eds) © 2006 Taylor & Francis Group, London, ISBN 0 415 40005 8

Geotechnical risk management in tunnelling

P. Schubert
iC consulenten Ziviltechniker GesmbH, Bergheim, Austria

ABSTRACT: Risk management in tunnelling is a continuous process from the alignment selection to the completion of the project. The usual steps of risk management, such as risk identification, avoidance or reduction during design development, risk sharing with the contractor, and controlling the remaining risk during construction equally apply to the tunnelling situation. Some basic elements of the risk management methodology and case histories are given.

1 INTRODUCTION

Risk is a steady companion of all construction people, who are dealing with natural ground. Nature is too complex and the means to investigate the true conditions are too limited to avoid surprises with confidence. If one understands risk as the, "possibility that a process changes to the worse", then risk is an inherent element of any underground design and construction.

Risk arises from unexpected conditions, either from rare and not sufficiently considered loads (e.g. earthquake) or – more frequently from unexpected properties of the ground. The tolerance of Clients, Engineers and Contractors to handle risk in a pragmatic manner is in a highly competitive environment small. The understanding of the unavoidable uncertainty in the ground is generally rare. Due to the raising awareness of safety and the need to make decision-making more transparent by thorough documentation, risk management has become an inherent element of dealing with difficult underground construction.

Risk management in a construction project can be understood as an aspect of a comprehensive quality management system. In this view it makes also sense to understand risk management as a continuous process of improvement during the project life. Risks should be identified and dealt with as early as possible and reduced or avoided by project adaptations or technical measures. The residual risks during construction require a methodology that ensures timely identification of unfavourable developments and implementation of countermeasures.

1.1 *General approach*

Risk management is being performed in numerous areas of science and business, particularly in the financial world for dealing with investment risk, in the chemical and nuclear industry where the consequences of failures would be disastrous, as well as in the areas of medicine or environment. Also in construction, risk management increasingly becomes an element of a comprehensive safety concept, e.g. at the design of dams (e.g. DIN 19700) (Rissler, P).

The fundamental approach is the same in all areas, and comprises the following steps:

- Risk identification.
- Assessment of probability of occurrence and potential consequences, either in life or monetary value.
- Risk reduction or avoidance through changes or measures.
- Risk management to control the residual (or unavoidable?) risk.

These steps can be applied to geotechnical problems without limitations.

1.2 *Project phases*

Geotechnical risk management takes place in all project phases (Figure 1). With the gain of detailed knowledge of the project, risk assessment is improved, from general geological questions at the project start to more construction – related questions towards the end of the design phase.
Phase 1: Alignment design and preliminary project

Typical risk management activities are:

- Identify general geological risk factors.
- Make a first risk classification, set priorities.
- Optimise alignment and/or location of construction.
- Assess and select basic construction methods.

Phase 2: Detailed design

- Improve risk assessment of the selected alignment and construction methods.
- Assess probabilities of occurrence and potential damage.
- Minimize risk through design adaptations or additional measures.
- Define the residual risk, the expected behaviour of the construction.
- Define the criteria for the observational method.

Phase 3: Tender phase

- Clarify risk ownership.
- Define methods to deal with uncertain elements commercially in a fair manner.
- Optionally include the Contractor into risk assessment.

Phase 4: Construction phase

- Develop safety management system.
- Specify and organise observational method.

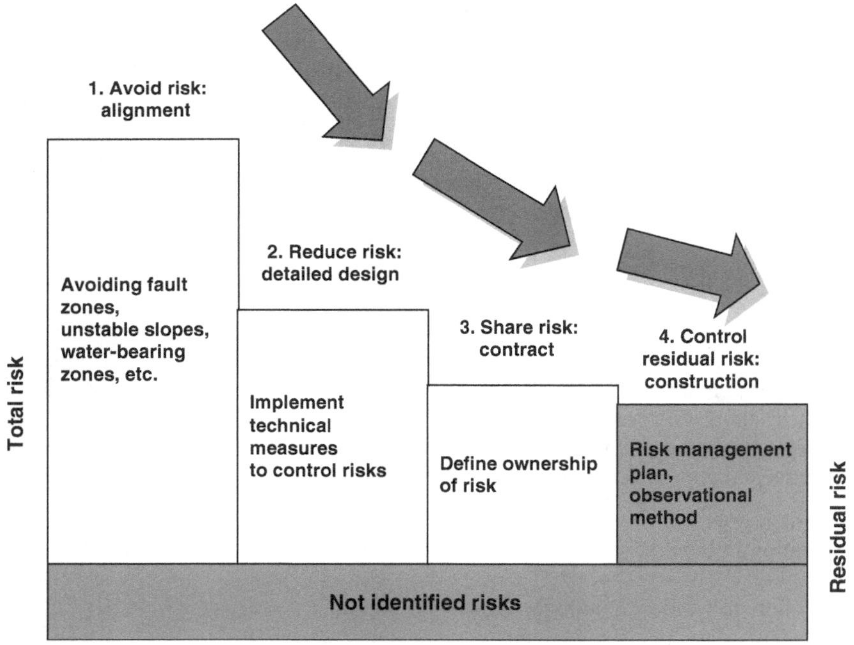

Figure 1. Risk management and risk reduction during the project phases.

2 RISK MANAGEMENT DURING DESIGN

At the beginning of the design phase the focus is risk identification. Risk identification must be continued during the design period, since additional knowledge will be available through further investigation and the specific construction concept. New risks may arise and previous risks may disappear. Typical geotechnical risks might be:

- Unstable slopes or rock falls at road or rail alignments or tunnel portals.
- Problems with construction through fault zones, low strength of the rock mass, lack of stability and squeezing conditions.
- Potential effects of the project to the environment, such as settlements or vibrations.
- Changes of the natural water regime, water inrush in tunnels.
- Carst cavities, etc.
- Earthquake loads
- etc.

It is useful to classify the risks according to their probability of occurrence and potential damage. In most cases, there is no basis to determine these parameters with accuracy. Therefore the subjective assessment of a group of experts and the use of ranges becomes important. A well documented assessment of experts is generally more appreciated than calculations only – at least in the geotechnical field.

The risk is defined as the product of probability of occurrence and the potential damage. We have recently applied the risk matrix shown in Figure 2 for tunnel projects. It was found necessary to define probabilities and damage specifically for each construction project and matrices from other fields cannot be directly applied.

In the matrix of Figure 2 the damage categories with a rating 0.5–3 are purely monetary criteria. Damage rating 1 deals with the risk of major changes in the quantity of specific items of the BOQ, which may in most countries lead to a claim. Damage rating 3 deals with the need to apply major measures that have not been foreseen in the contract and are therefore mostly very expensive.

Potential damage

Probability		10	5	3	1	0.5
		Accident life and material damage	Accident material damage	Exceptional measures	Regular measures	Regular measures
		Damage affects also third party	Damage limited to project	Not foreseen in the contract	Risk of excess quantities at specific items	High confidence of prognosis
Frequent	10	**100**	**50**	**30**	**10**	**5**
Occasional	5	**50**	**25**	**15**	**5**	**2.5**
Random	2	**20**	**10**	**6**	**2**	**1**
Seldom	1	**10**	**5**	**3**	**1**	**0.5**
Remote	0.5	**5**	**2.5**	**1.5**	**0.5**	**0.25**

Risk rating

>20	**Not acceptable, design adjustment necessary**
10 – 19	**Critical, special measures required in the design**
5 – 9	**Acceptable remainig risk, risk management required**
0 – 4	**Low remaining risk**

Probability

Frequent	**May occur several times in the project**
Occasional	**May occur once in the project**
Random	**Unexpected in this project, but occurs sometimes in similar projects**
Seldom	**Unexpected in this project, but occurs seldom in similar projects**
Remote	**Expert group considers this event unprobable to occur in this project**

Figure 2. Example of a risk matrix applied to a tunnel project in Austria.

The damage categories with rating 5 and 10 deal with accidents, unforeseen major events. Rating 5 has to do with damage to property, or monetary damage to repair or reconstruct, but limited to the parties of the construction project. Rating 10 includes personal injury, mostly in addition to monetary damage or major damage to property that includes public interest.

We have found that – compared to other fields – the probability of occurrence for the usual risks of a tunnelling project are relatively high and are also accepted by the tunnelling community. Events with a very low probability of occurrence can hardly be assessed with confidence due to the lack of reliable statistical data and the difficulty to compare the boundary conditions of different projects.

At the moment, quantitative risk assessments using probabilistic methods (e.g. Monte-Carlo method) must be treated with care. Experience showed that monetary consequences are mostly underestimated and thus in contradiction with actual cost increases of tunnel projects.

A logic has been established how to deal with risks that could not be reduced to an acceptably low level à priori (Figure 3). This logic is based on the observational method according to Eurocode 7. We consider this approach adequate for critical projects, such as tunnels with low overburden, rather than for deep tunnels. In deep tunnels, with much more un-knowns, the observational approach is traditionally applied in a much more general approach with less strict requirements. The main problem there is, that the risk may remain unknown until it appears, and can thus not be managed through a suitable plan.

The steps in this flowchart are the following:

- Initially, the risk must be assessed reliably through design calculations or precedent experience in comparable conditions. If either of these can be answered positively, the observational approach of EC7 can be applied. Otherwise the design needs to be changed.
- The observational part (yellow) has two critical elements:
  – Can the risk be identified in due time to allow an adequate reaction?
  – Are practicable methods readily available to counteract unfavourable developments?
- Both elements need to be answered positively to allow this observational approach at tunnels with low overburden. If one of the two criteria is negative, the project would need to be changed.

3 RISK MANAGEMENT DURING CONSTRUCTION

At the beginning of construction usually a number of residual risks are left over as a result of the uncertainties of the natural ground, including known risks that cannot be avoided or risks that have not yet been identified. Depending on the legal environment of the project and the conditions of the contract, the Client and the Contractor share the risks. Usually the ground conditions remain the risk allocated to the Client, and the Contractor's risk is mainly the construction method and the use of the resources.

Today's requirements to deal with safety and the need to make decisions transparent suggest to establish a risk (or safety-) management plan as an element of the project quality management system before construction. In the perspective of the Client, whose main concern is the stability of the tunnel, the main elements of the safety management plan are:

- The definition of the expected behaviour of the tunnel (settlements, convergence, surface settlements, etc.) and the tolerance limits.
- The set-up of an adequate organization and technology to monitor the behaviour of the tunnel including the main factors of influence, such as geological documentation, prognosis, follow-up of the geotechnical model and their interpretation.
- The comparison of expected and actual behaviour and the adjustments with regard to experience made (this will change with the learning curve of every project).
- The well documented reaction to deviations from the expected behaviour.
- Alarm criteria, organization, priorities and measures in case of a crisis.

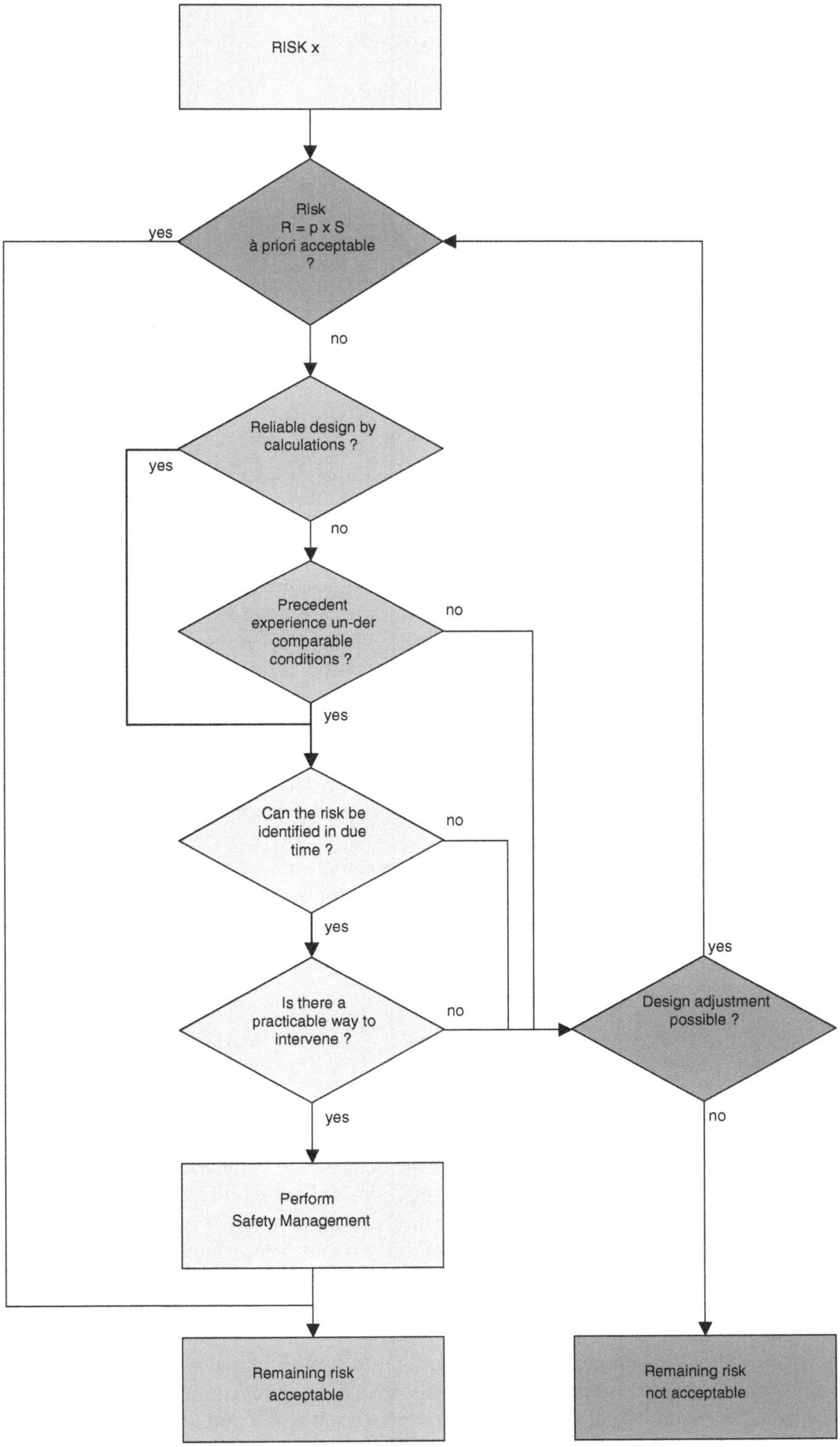

Figure 3. Logic of dealing with residual risk, not avoidable à priori.

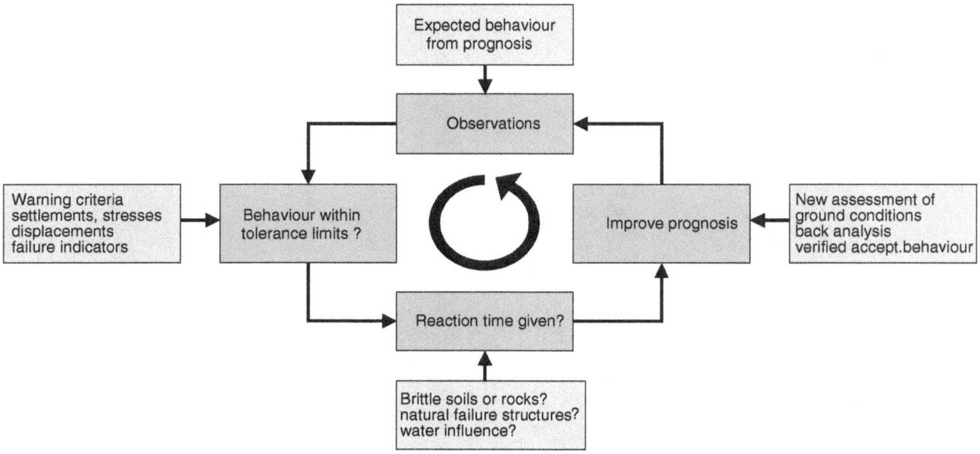

Figure 4. Risk management during construction following the observational method according to EC 7.

At this point the NATM approach with the determination of the support based on certain obserrvational elements, Pacher et al. (1974) and the anglo-american, more formal approach of the "observational method", Peck (1969) and the version in Eurocode 7 (1997) meet.

As mentioned above, a suitable reaction time and the follow-up of the expected behaviour (incl. tolerance limits) are two very demanding elements of the observational method. Experience shows, that the tolerance limits established during design are often violated in execution, but sufficient other elements of assessment are available to confirm an acceptable level of safety. For this purpose direct stress (strain) measurements, back analyses, new data on the ground properties, and statistical analysis of recent monitoring data can be very useful. Therefore the safety management plan must allow for the implementation of new findings in terms of a permanent follow-up.

Using the approach given in Figure 4 not only the actual safety level of a tunnel construction can be enhanced, but also a formally correct and sufficient confirmation of the safety level can be achieved according to EC 7.

4 CASE HISTORIES

4.1 *Case history underground museum in Salzburg – concept design*

The underground Mönchsberg Museum in Salzburg is a concept of Hans Hollein that has been discussed since 1990. The design is based on an underground complex with exhibition halls, depots, shops, infrastructure and supply shafts. The core of the complex is a huge conical shaft with a diameter of 30 m at the top. The concept of Hans Hollein requires the conglomerate walls of the shaft to remain unsupported and in the natural appearance of the rock.

The first feasibility study was made in 1990, but no subsurface geotechnical investigations were undertaken. In 2001 additional funding was provided to sink two core borings in the critical zone of the project and several trenches. Laboratory tests were made to obtain mechanical parameters of the rocks.

The new study should – amongst others – highlight the geotechnical risks, which had received considerable contingencies in the cost estimate of 1990, because of the uncertainty if the natural rock surfaces in the central shaft would cause water seepage and require support.

The whole complex lies in the Mönchsberg, the natural Western boundary of the historical town of Salzburg. The Mönchsberg is the rest of a post-glacial gravel delta, which had been cemented to conglomerate. The following geotechnical risk scenarios were identified:

- Locally poor or complete lack of cementation of the conglomerate and insufficient strength. Besides the difficulties to excavate such material, the efforts for stabilisation and surface treatment were of concern.
- Water seepage from the open rock surfaces, violation of the extremely strict requirements of humidity control in the museum.
- Unfavourable intersection of large joints or faults with the caverns requiring extensive support, such as pre-stressed anchors.
- Natural or artificial caves at the boundaries of the openings that might require additional support or filling.

New solutions were sought during the risk assessment to reduce the consequences of the some risk scenarios. The most interesting new concept was developed to deal with the problem of water seepage into the central shaft. The concept includes a small base tunnel and a cut slot around the shaft. The cutting should be performed with a rope saw (easy in the conglomerate) and the slot equipped with a waterproofing membrane. By these means water access to the shaft could be completely avoided at very reasonable and well defined cost. The potential cost of the risk factors are listed in Table 1.

The occurrence and cost data were processed in a Monte-Carlo analysis. The estimate of the likely cost of the geotechnical risk factors could be much improved compared to the 1990 study and the contingency budget considerably reduced.

4.2 *Case history Lainzer Tunnel Lot LT22*

The Lainzer Tunnel is a 12.8 km long connection of the railway lines Westbahn, the Südbahn and the Donauländebahn in the City of Vienna. Lot 22 consists of two single track tunnels under-passing a hill and the Westbahn. The length of the tunnels are 750 m and 900 m respectively, the soil cover 10 m in average. Both tubes were built using the NATM concept.

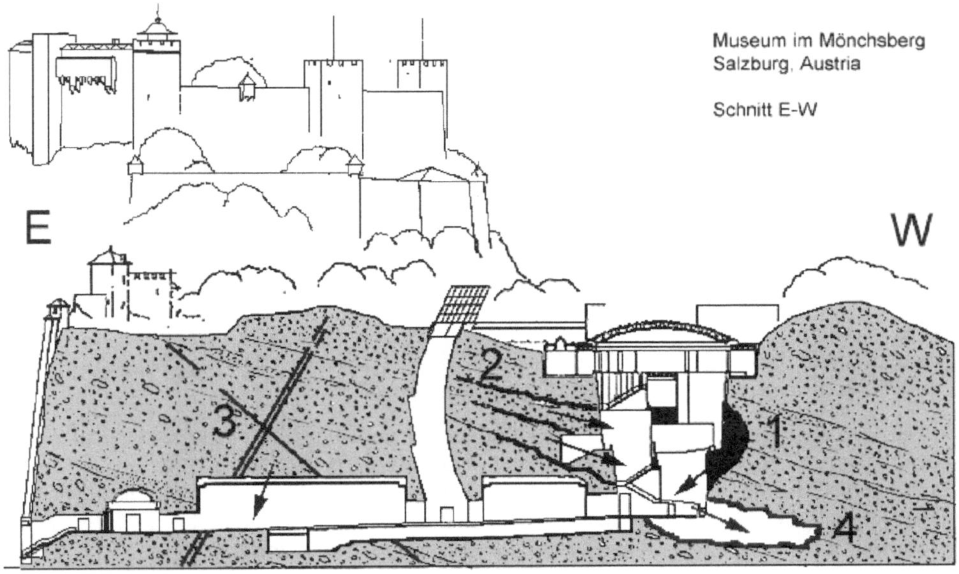

Figure 5. E-W section through the museum complex with geotechnical risk factors: 1. unconsolidated rocks, 2. water ingress, 3. wedge failure at geological structures and 4. natural caves.

Table 1. Input data for the geotechnical risk analysis: risk factors from geology and geotechnics.

Geotechnical risk factors	Probability of occurrence From	To	Quantity	Unit
R1 loose, un-cemented conglomerate	1%	3%	96.000	m^3
R2 need of grouting to strengthen the rock mass	5%	10%	12.000	m^3
R3 need of surface treatment of the conglomerate	30%	70%	3.000	m^2
R4 extensive water seepage	0%	2%	96.000	m^3
R5 major unfavourable joint, need of pre-stressed anchors	50%	150%	10	Stk
R6 treatment of caves	110%	130%	1.000	m^3

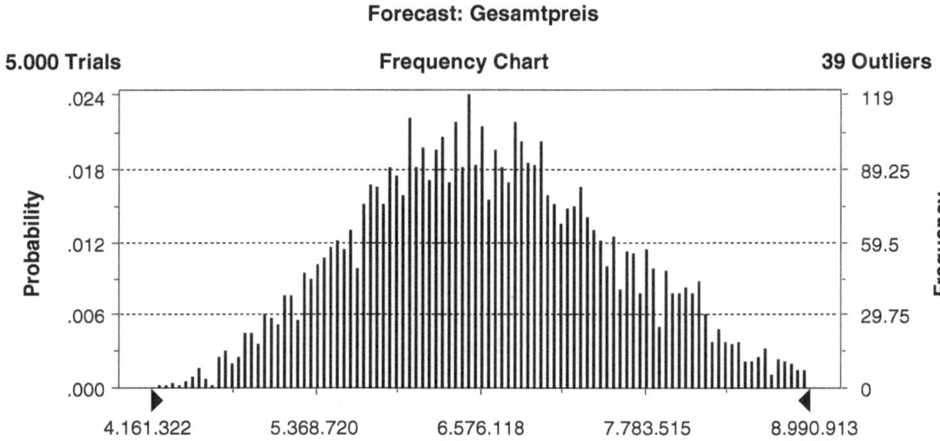

Figure 6. Expected variation of cost from the geotechnical risk factors.

Based on the so-called framework-plan, commonly applied at the projects of the Eisenbahn-Hochleistungsstrecken AG, a geotechnical safety-and-crisis-management-plan was developed. This plan included the following items:

- Responsibility of all parties involved in the safety management.
- Expected behaviour of the tunnel and tolerance limits.
- Monitoring program, frequency, information flow.
- Geotechnical meetings.
- Critical scenarios, warning levels.
- Procedures and organisation in case of crisis.
- Preparation of material and equipment for the case of crisis.

Vavrovsky et al. (2001) have reported earlier about the details of this system. Practically most important is the clear and simple daily reporting and warning system.

In this project the daily information was compiled to a table including all monitoring sections. At first sight this table shows those sections with unusual behaviour and the warning level. In case of a warning level being reached, an explanation and remedial measures are given (Table 2).

In reality, where variable ground conditions prevail, the prognosis and warning levels will be frequently violated. Nevertheless the total amount of monitored data will usually allow a good assessment of stability, mostly more favourable than foreseen. This is why a number of qualitative warning criteria were foreseen in this project besides the quantitative data (Table 3).

Table 2. Daily geotechnical report, extract.

Tunnel	Section	Behaviour/warning level Normal (0)	Deviation (1)	Critical (2/3)	Geotechn. explanation see ( )	Measures required
BW7	MQ237-UT	✓				
	MQ246-UT	✓				
	MQ255-UT	✓				
	MQ264-UT		!		(2)	Enhanced support of the tunnel face
BL7	MQ138-UT	✓				
	MQ147-UT		!		(1)	Check strength development of shotcrete
	MQ156-UT		!		(1)	
	MQ191-UT	✓				
	MQ200-UT	✓				
	...					
	...					
	...					
	(1)	geotechnical explanation 1:...				
	(2)	geotechnical explanation 2:...				

Table 3. Levels of the geotechnical warning system.

Warning level 0	**System behaviour within expected behaviour** Normal behaviour according to the forecast, no warning criteria violated, no particular observations
Warning level 1	**Deviation from normal behaviour, violation of a warning criterion** A warning criterion could be: Exceeding a warning value given by the design atypical displacement histories of the tunnel and/or the surface, such as sudden additional displacements, unusual changes in the trend line, failure indicators a tunnel linings, etc. smaller unforeseen events in the tunnel, over-break, major water ingress, etc.
Warning level 2	**Imminent danger, risk limited to job site** Alarm criterion violated, however not in the influence zone of the railways. The alarm criterion can be: imminent collapse of the tunnel violation of railway tolerance values repeated progressive tendencies of tunnel displacements persistent doubt on the stability of the tunnel, signs of stress, cracks, etc. flooding of the tunnel
Warning level 3	**Imminent danger, risk not limited to job site, public interests affected** Alarm criterion violated with potential effect on the railways

The experience of the project showed that some additional aspects should be considered at future tunnel projects with low overburden:

- The difficulty to correctly correlate the predicted geotechnical model to the actual conditions and locations in a project with variable ground conditions, such as the Lainzer Tunnel LT22. Consequently the problem to define the "expected behaviour". At this project it would have been better and easier to have a pattern of potential ground conditions with corresponding tolerances and solutions, instead of a pre-determined allocation of certain ground conditions. The relevant type of ground conditions should be identified through the geological and monitoring data.

- Priorities should be more clearly defined in case of crisis.
- A better coordination between the various plans for risk an crisis management between the job site and public institutions.

5 CONCLUSION

Geotechnical risk management in tunnels does not require any particularly difficult methods or technologies. However, risk management: a) requires technical expertise in the relevant field and b) it has to be done. Following a few principles, risk can be minimised during all project phases and lead to a much improved transparency of decision making.

REFERENCES

Eurocode 7. Entwurf, Berechnung und Bemessung in der Geotechnik, ÖNORM ENV 1997–1.

Naval Surface Warfare Center Carderock Division, Standard Operating Procedure. www.dt.navy.mil/env/code 0073sop.html.

ÖNORM EN 1050 1.Jänner (1997). Sicherheit von Maschinen, Leitsätze zur Risikobeurteilung

Pacher, F., Rabcewicz, L.V., and Golser, J. (1974). Zum derzeitigen Stand der Gebirgsklassifizierung im Stollen- und Tunnelbau. Proc XXII Geomechanik Kolloquium Salzburg pp. 51–58.

Peck, R.B. (1969). Advantages ad limitations of the observational method in applied soil mechanics, *Geotechnique* 19, no. 2.

Rissler, P. Dimensioning of the design flood as part of a reservoir safety concept, Internetseite des Deutschen Talsperren Komitees (DTK) www.germannatcom-icold.de

Vavrovsky, G.M., Ayaydin, N., and Schubert, P. (2001). Geotechnisches Sicherheitsmanagement im oberflächennahen Tunnelbau, *Felsbau* 19 no. 5.

Geotechnical Risks in Rock Tunnels – Campos e Matos, Ribeiro e Sousa, Kleberger & Lopes Pinto (eds) © 2006 Taylor & Francis Group, London, ISBN 0 415 40005 8

Use of decision aids for tunnelling

Herbert H. Einstein
MIT – Massassuchets Institute of Technology, Cambridge, MA, USA

ABSTRACT: This paper is based on the oral presentation with the same title, given at the Aveiro Course on Geotechnical Risk in Tunnels in 2004. The main part of this paper consists of a review of recently published material on the DAT (Decision Aids for Tunnelling) which give a general description and show the application in the Swiss Transalpine tunnels both before and during construction as well as an application to assess different tunneling technologies. The possibility to update the information and resource management are also discussed in this context. Finally, two cases in Korea are presented in which the effect of exploration and of different construction methods are described.

1 INTRODUCTION

This paper is an abbreviated written version of the oral presentation made at the workshop, Geotechnical Risk In Rock Tunnels, Aveiro, Portugal, April 16–17, 2004. It is an abbreviated version since The Decision Aids for Tunnelling (DAT) have been described in a number of publications (e.g. Einstein et al., 1996, Xu et al., 1998, Descoeudres et al., 2000) which can be easily accessed. The present paper will, therefore, in Section 2, briefly summarize the relevant information described in some of these earlier publications and which was orally presented at the Aveiro workshop. The material which was new when presented at the workshop is the topic of Section 3. Section 4 contains the conclusions.

2 RISK, SUMMARY OF PREVIOUSLY PUBLISHED MATERIAL

2.1 *"Decision Aids for Tunnelling", Transportation research record, No. 1656 (Einstein et al., 1999)*

This paper starts with a general description of the DAT, and then discussed applications to Swiss transalpine tunnels and to an investigation of the effect of different construction methods for freight tubes under cities:

The DAT consists of two types of components:

Description of Geology
Construction Simulation

The geologic description uses geologic/geotechnical descriptions of the ground including the associated uncertainties to produce geologic/geotechnical profiles as shown in Figure 1. The differences between the resulting ground class profiles in the figure express the uncertainty, and by simulating (with Monte Carlo Simulation) a large number of such profiles the overall geologic/geotechnical uncertainty is expressed.

Construction is then simulated through each of the ground class profiles with each ground class being associated with a particular construction method (Figure 2). Variability in construction performance is expressed through advance rate- and cost distribution (also shown in Figure 2).

Combining geologic and construction uncertainty in simulating produces the cost-time scattergram shown in Figure 3.

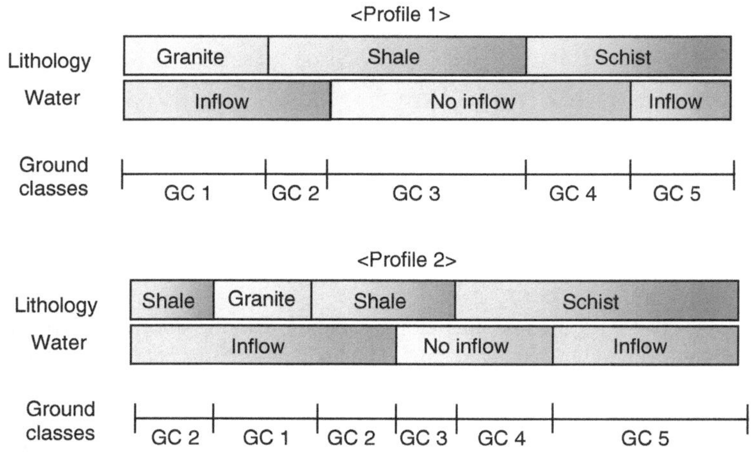

Figure 1. Geologic/Geotechnical (Ground Class) profiles expressing uncertainties. Geologic parameters lithology and water are uncertain as expressed by different lengths. The resulting ground class profiles are correspondingly different.

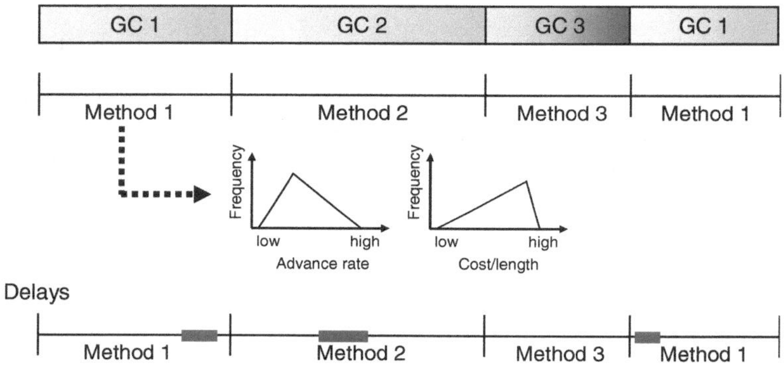

Figure 2. Specific construction methods are associated with each ground class. Construction methods vary in cost and time even if the geology (Ground Class) is constant.

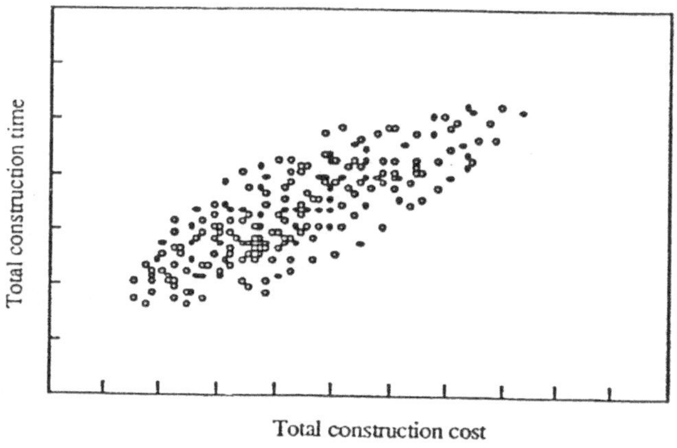

Figure 3. Resulting time-cost scattergram.

The DAT have been applied to the Swiss Transalpine Base tunnels, which are being built at the present time, the Gotthard and Lötschberg Base tunnels. Figure 4 shows the scattergram results of the initial investigation during the planning phase (early 1990's) when the different tunnels systems shown in Figure 4 (Double Track plus service tunnel, 2 single track plus service tunnel, 3 single track tunnels) were investigated. The three separate "clouds" for each system represent the effect of different lengths (0, 0–20, 20–50 meters) in the problematic Piora Zone. The length effect of the Piora Zone was investigated further by looking at greater lengths up to 400 m (Figure 5).

The TRR No. 1656 then continues with the application of the DAT to assess different tunneling methods and, particularly, the effect of improvements in these methods when constructing a network of freight tubes under cities. The tunneling methods are pipe jacking and micro-tunnelling for tunnel diameters of 1.2 and 2.4 m and TBM for tunnel sizes of 4.3 m. In applying these methods to corresponding tube networks (Figure 6) and examining the effect of technical improvements, the results shown in Figure 7 are obtained.

2.2 *"The Decision Aids for Tunnelling (DAT) – an update" transportation research record no. 4850 (Einstein, 2004)*

After a brief description of the DAT and some applications (not repeated here since they correspond to those above) this paper discusses updating and resource management using the DAT:

Updating which has been discussed in great detail in another paper (Haas and Einstein, 2002) uses geologic and construction information obtained during construction to revise the cost and time predictions for the as yet unexcavated part of the tunnel. This is shown in Figure 8 distinguishing two possibilities: 1) Reducing the spread (uncertainty) of the time-distance diagram by replacing the

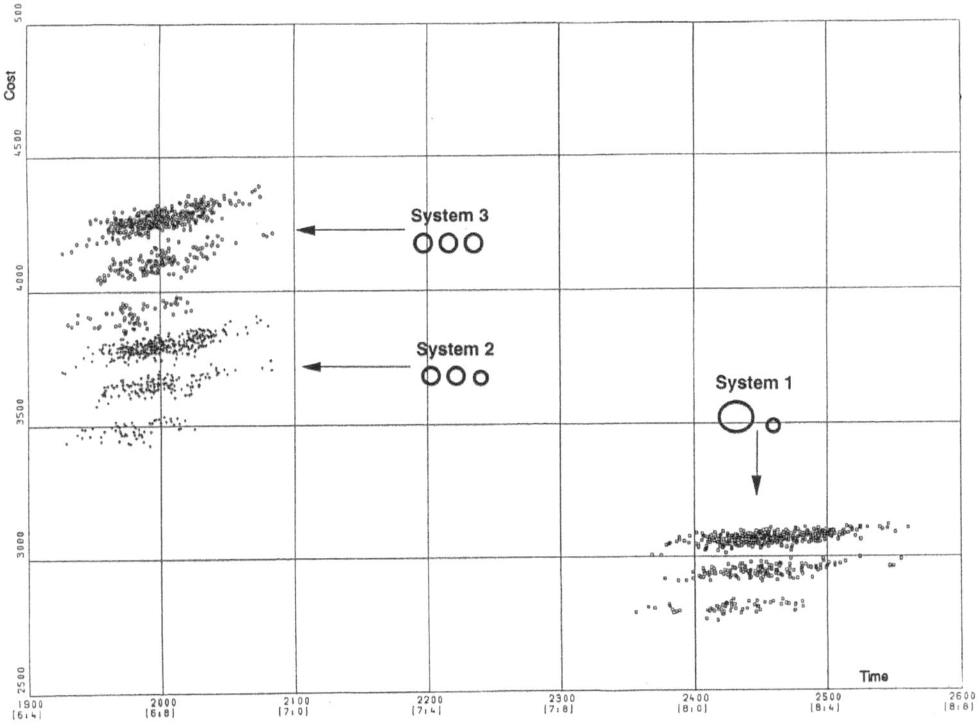

Figure 4. Scattergrams for Gotthard Base tunnel, investigation of cost and time of three different tunnel systems in planning phase.

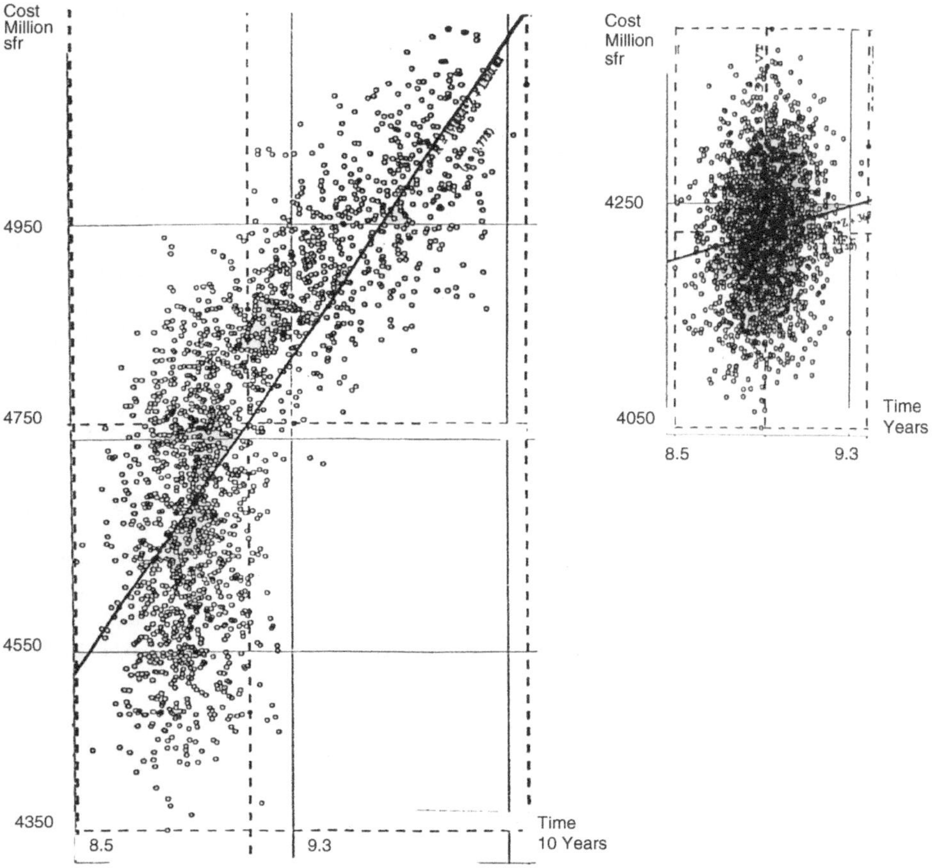

Figure 5. Cost time scattergrams – Gotthard Base tunnel. The left figure shows cost increase but no time increase for lengths of Piora Zone below 200 m, followed by cost and time increases for lengths above 200 m.

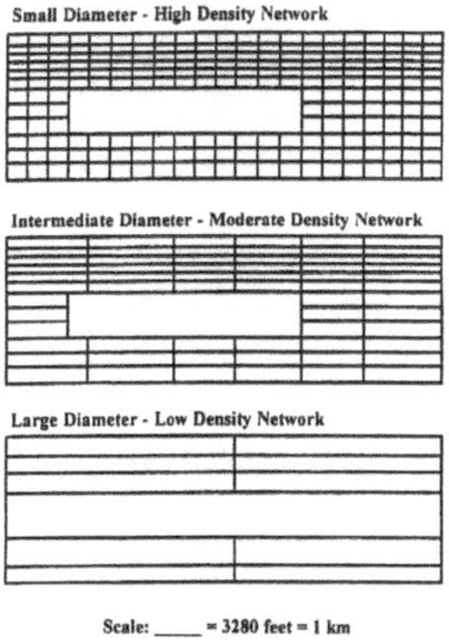

3 diameters (1.2 m, 2.4 m, 4.3 m) applied to different existing scenario e.g. midtown Manhattan

Different Construction Technologies: Small: Micro Tunnelling
Large: TBM

Figure 6. Assessment of different construction methods freight tube systems in cities.

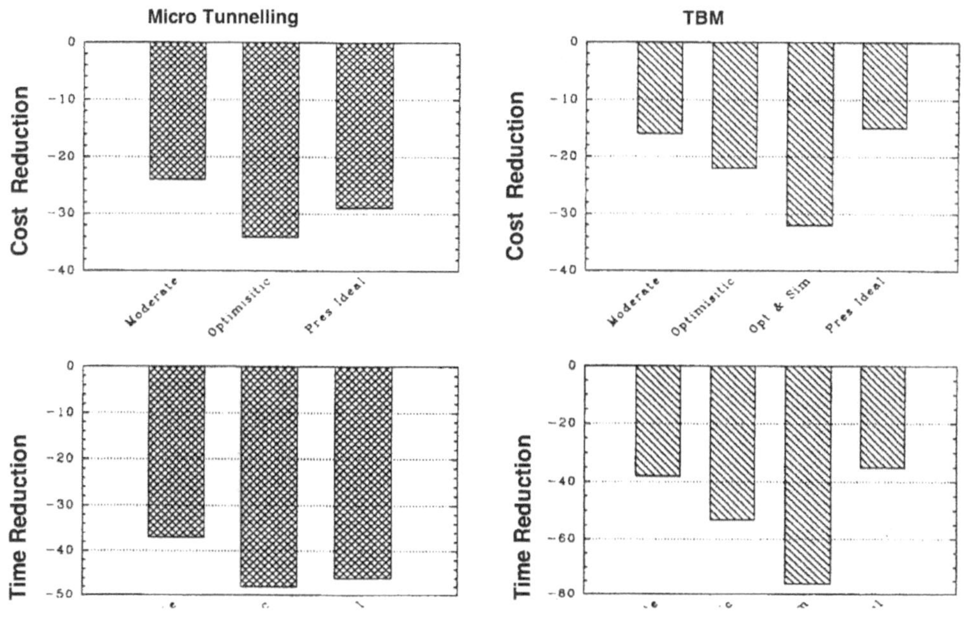

Figure 7. Assessment of different construction methods different potential improvements of construction technology. Cost and time reduction in %.

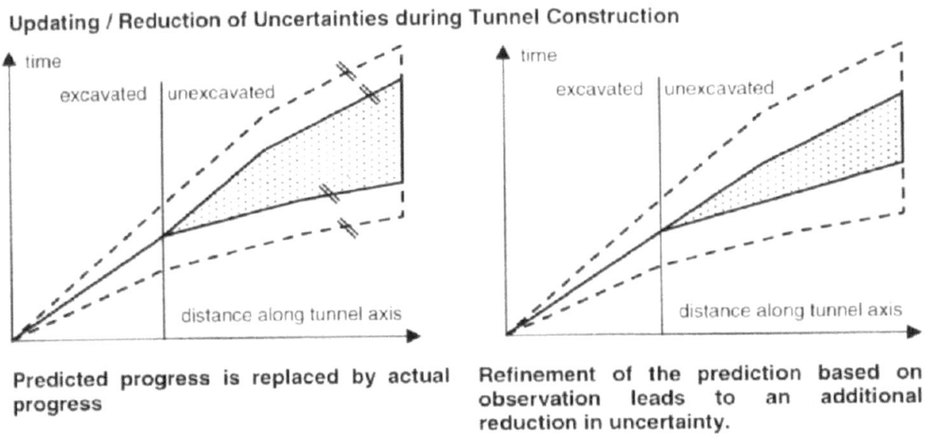

Figure 8. Updating principle.

predicted progress with the actual one; 2) Learning from experience in the excavated part to reduce the uncertainty (spread) in the unexcavated part. The effect of the two different types of updating can also be shown in cost-time scattergams (Figure 9). While Figure 8 and 9 illustrate a "synthetic" case, updating was also used in the actual case of the Lötschberg Base tunnel and Figure 10 shows time distance diagrams with spreads developed at different dates. The time distance diagrams b) and c) reflect actual progress and the experience gained in the excavated part of the tunnel.

The DAT can also be used for resource management. Required resources such as materials, manpower and equipment and produced resources such as muck and water are affected by the

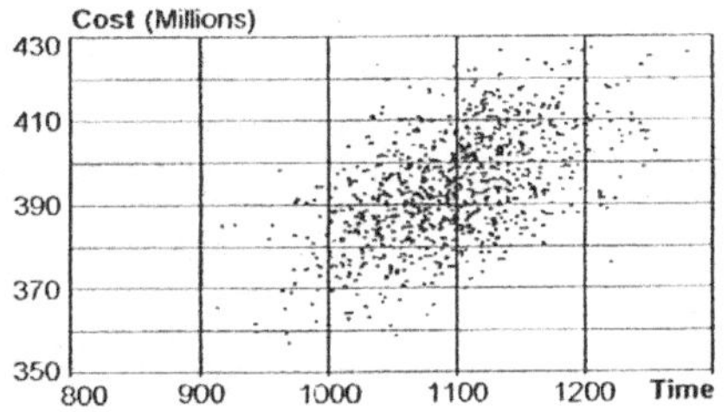

Time–cost scattergram for initial input - tunnel before construction starts

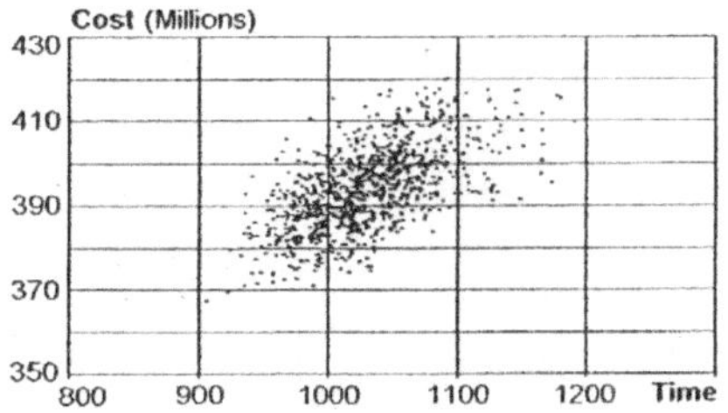

Time–cost scattergram for partially updated input tunnel after some excavation has taken place use only observed information in excavated part

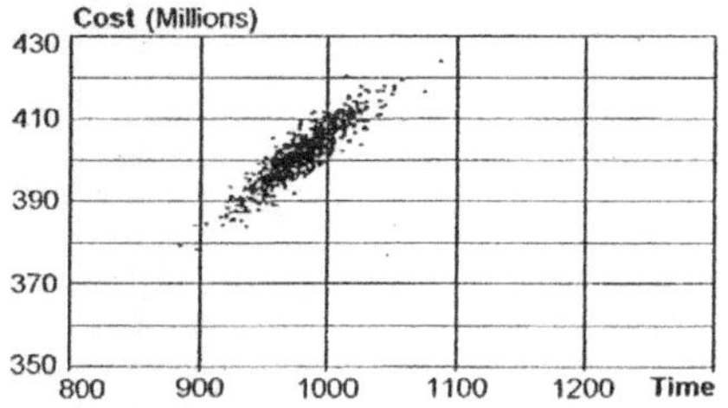

Time–cost scattergram for fully updated input - tunnel after some excavation has taken place; observed information in excavated part used to update unexcavated part.

Figure 9. Updating a synthetic tunnel – effect on cost–time scattergrams.

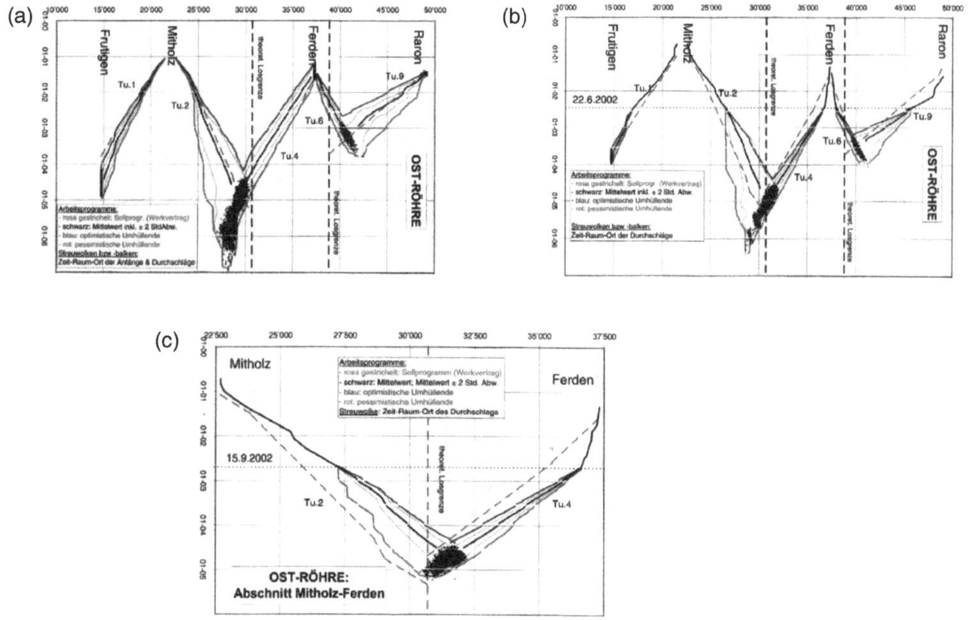

Figure 10. Updating for Lötschberg Base tunnel.

a) Time Distance Diagram May 2002
b) Time Distance Diagram June 2002
c) Time Distance Diagrams September 2002 (Shown for critical section only)
Note: Reduced Uncertainty from a to b to c.

tunnelling process and the associated construction- and geologic uncertainties for instance:

- Muck characteristics depend on geology and related uncertainties, and on the excavation process.
- Muck volume per time depends on advance rate which in turn depends on geology/construction process and related uncertainties.
- Muck characteristics will vary even for a constant geology and a given excavation process.

In the Lötschberg Base tunnel in Switzerland, muck is to be reused to produce concrete aggregate. This raises the following specific issues:

Muck is to be reused as much as possible for concrete or shotcrete aggregate.

TBM muck can be used for shotcrete but not concrete.

Added complication: alkali-reactive material cannot be used as aggregate or requires additives.

Material that cannot be used for shotcrete/concrete needs to be disposed of.

If not enough aggregate for shotcrete/concrete can be produced from muck, additional quantities need to be brought in (bought).

Since the type and quality of muck is subject to uncertainty, as mentioned above, the type and quantity of different resulting materials is subject to uncertainties. When managing resources, it is important to have an idea about these uncertainties and this can be done with the DAT. A study was conducted at MIT to demonstrate the application of the DAT for resource management using the Lötschberg Base tunnel as an example. From a geologic point of view, the muck categories K1, K2, K3 (K1 best) can be distinguished while construction is either by TBM or drilling and blasting (DB). Considering the geology and construction methods and the muck related issues listed above, this leads to the following muck categories:

- K1DB and K1TBM: K1DB used for cast in place concrete, K1TBM useable for shotcrete.
- K1arDB and K1arTBM: Same usages as above but only with special mix because of alkali reaction (ar).
- K2DB and K2TBM: Used if necessary for shotcrete and cast in place concrete
- KuTBM and KuDB: This includes alkali reactive K2 material produced by TBM or DB and all K3 material independent of the method of excavation. Needs to be disposed of eg. in landfills.

The muck is excavated in different tunnel sections and stored in different repositories. From there it is transported to the appropriate plant where it is sorted into different grain size categories and used for concrete fabrication or it is transported to disposal sites. All this is shown in Figure 11. With the DAT one can then get an idea how different material types (Figure 12) or aggregate components (Figure 13) accumulate or diminish with time (construction process). The plots allow the decision maker to know when there is a surplus or need for additional material; they can include uncertainty similar to the time distance diagram.

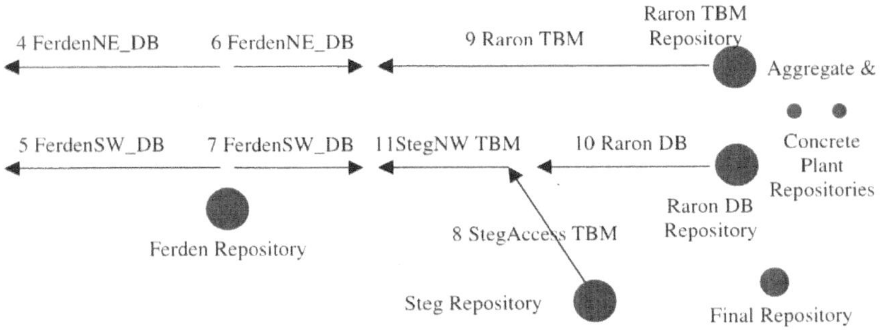

Figure 11. Materials management Lötschberg Base tunnel.

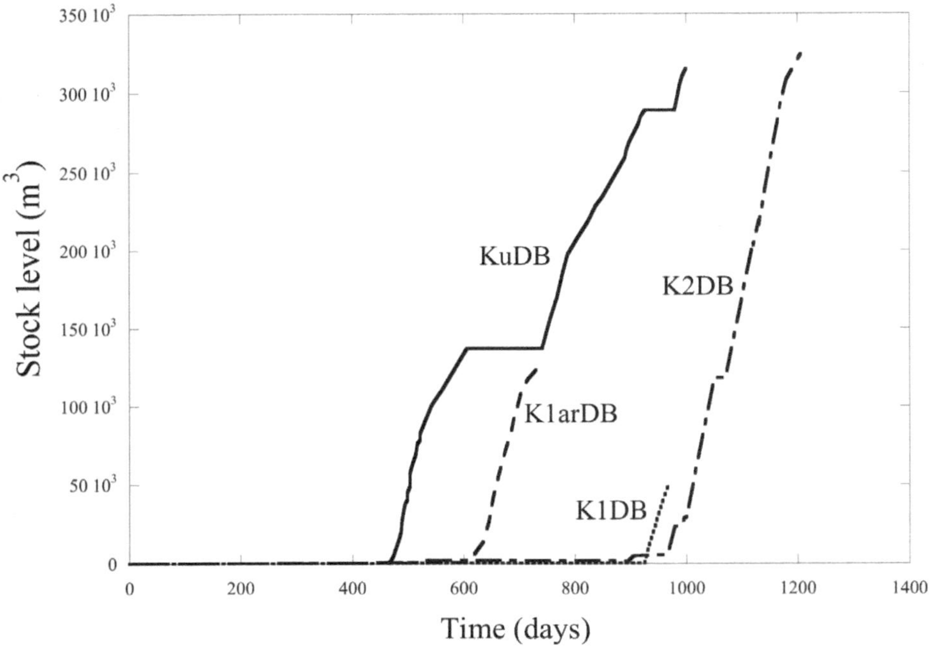

Figure 12. Resource management. Drill & blast material in Raron repository of the Lötschberg Base tunnel.

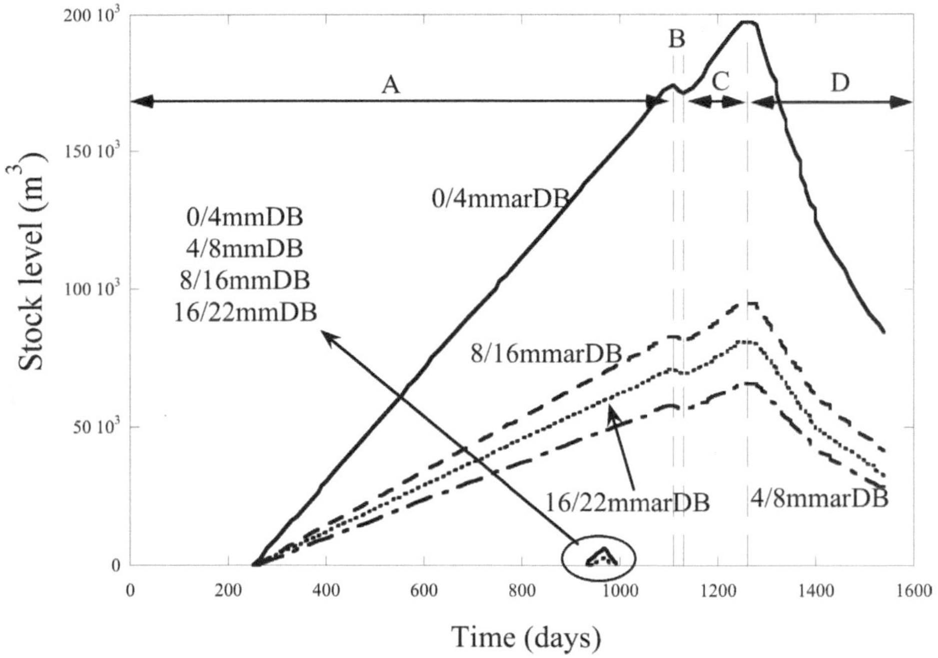

Figure 13. Resource management. Different concrete aggregate components in Raron repository of the Lötschberg Base tunnel.

3 DIFFERENT DAT APPLICATIONS

In this section, applications are described which were presented at the Aveiro workshop and were not published earlier or published in a less accessible context.

3.1 *Effect of exploration*

Exploration usually reduces uncertainty. This has been investigated using the DAT by Descoudres and Dudt, 1999 and is shown in Figure 14. The figure shows the cost-time scattergram for the Gotthard Base Tunnel before and after consideration of borings in the so-called Tavetsch Intermediate Massive. This geologic formation is highly disturbed and exploratory borings were drilled to get a better idea on material properties. The uncertainty after exploration is less than before, as expected.

In this context, it is interesting to look at another case, the Wonhyo Railroad Tunnel in Korea where the DAT were applied to assess time and cost uncertainties. After doing a normal analysis, the case was used to artificially suppress geologic and construction uncertainty or both. All this is shown in Figures 15 and 16. In this case, geologic uncertainty affects mostly cost while construction uncertainty affects time. It should be noted that no general conclusions should be drawn from these results; they may be related to the specifics of this tunnel.

The effect of exploration was also investigated in the context of the planned Maurienne-St. Ambin Base Tunnel between, Lyon and Turin. Different length exploration galleries and the construction of

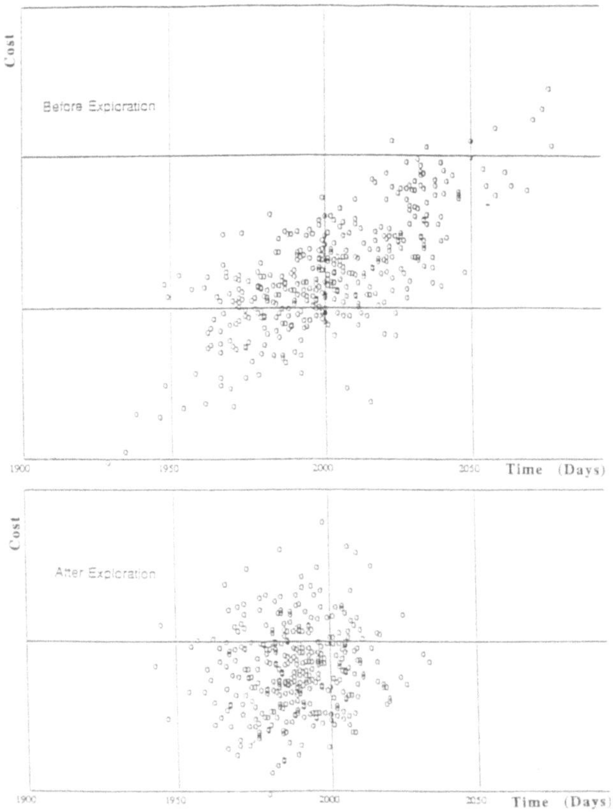

Figure 14. Gotthard Base tunnel. Effect of exploration in Tavetsch.

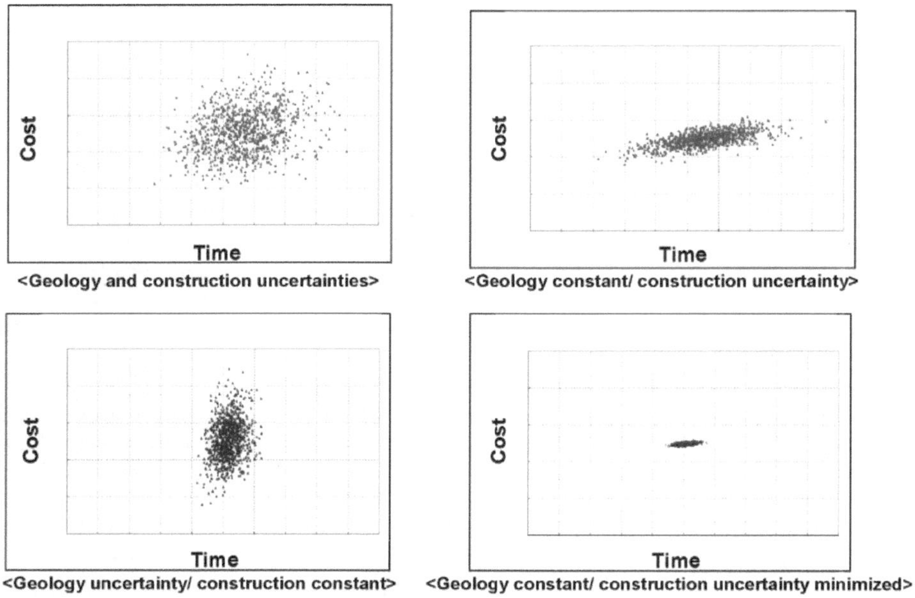

Figure 15. Effects of the major sources of uncertainty Wonhyo (Korea) railroad tunnel.

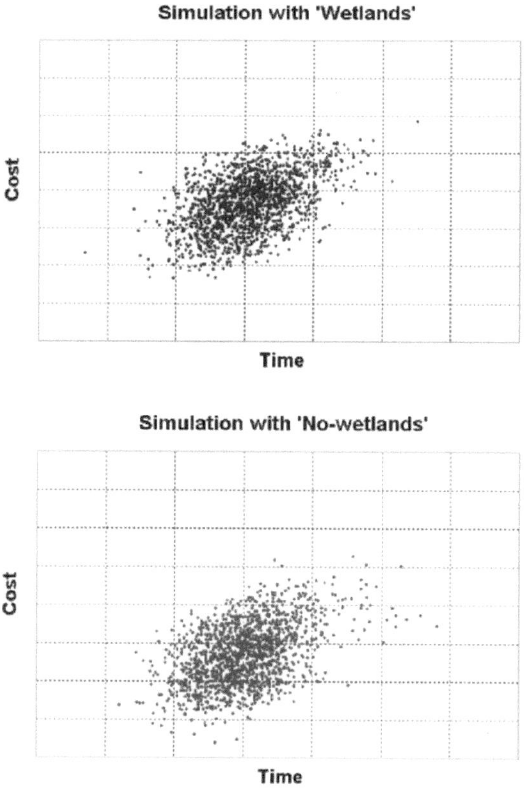

Figure 16. Effects of the environmental factor (Wetlands) Wonhyo (Korea) railroad tunnel.

these galleries prior to main tunnel construction or preceding main tunnel construction by a short time interval only, were investigated. The main effect of exploration tunnels is a reduction in time to build the main tunnel ranging from 10 to 40% while cost is only slightly reduced.

3.2 Effect of different construction methods

This, in essence, has already been discussed for the Gotthard Base tunnel where different lengths of "bad geologic conditions" were encountered (see Section 2.1). An analogous newer case is again the Wonhyo Tunnel in Korea (Figure 15). Construction "under wetlands" requires pre-grouting. This leads to an increase in cost but since grouting is done prior to tunneling, it does not affect the time.

4 CONCLUSIONS

The applications of the DAT presented in this paper and the ones discussed elsewhere lead to the following conclusions:

The DAT can consider most factors affecting tunnelling to determine construction cost and time as well as used and produced resources, all this with associated uncertainties.

Since the DAT consider uncertainties, the results provide a basis for risk analysis and modern decision making.

It is possible to include the effect of additional information when working with uncertainties.

REFERENCES

Descoeudres, F. and Dudt, J.-P. 1994. Instruments d'aide á la décision pour la construction de tunnels (ADCT). Publication No. 155 of The Swiss Society for Soil and Rock Mechanics.

Descoeudres, F., Dudt, J.-P., Einstein, H.H. and Egger, P. 2000. Instruments d'Aide a la Decision pour la Construction de Tunnels Appliqués a la Comparaison D'offres D'Entreprises, *Proc.* 2nd Int'l. Conf. on Decision Making in Urban and Civil Engineering Lyon.

Einstein, H.H. 2001. The Decision Aids for Tunnelling (DAT) – a brief review, *Korean Tunnelling Technology*, Vol. 3, No. 3, pp. 37–49.

Einstein, H.H 2004. Decision Aids for Tunneling – an update, Transportation Research Board, No. 4850.

Einstein, H.H., Dudt, J.-P., Halabe, V.B. and Descoeudres, F. 1992. Decision Aids in Tunneling; principle and practical application, *Monograph*, prepared for the Swiss Federal Office of Transportation.

Einstein, H.H., Dudt, J.-P., Halabe, V.B. and Descoeudres, F. 1996. Geologic uncertainty in tunnelling. *Proc*, ASCE Geotechnical Engineering Division Specialty Conference, "Uncertainty in the geologic environment: from theory to practice".

Einstein, H.H., Indermitte, C.A., Sinfield, J.V., Descoeudres, F. and Dudt, J.-P. 1999. The Decision Aids for Tunnelling, *Transportation Research Record*, No. 1656.

Einstein, H.H., Min, S.Y., Lee, J.S. and Kim, T.K. 2003. Application of Decision Aids for Tunneling (DAT) to a drill & blast tunnel, *KSCE (Korean Society of Civil Engineers) Journal of Eng.*, Vol. 7, No. 5, pp. 619–628.

EPFL. 1995. Finne Tunnel-Probabilistische Ermittlung der Baukosten und -Zeiten, Lab de Mec. Des Roches, ISRM, Report RX335.

Haas, C. and Einstein, H.H. 2002. Updating the Decision Aids for Tunneling, *ASCE Journal of Construction Engineeering and Management*, Vol. 128, No. 1, pp 40–48, Jan/Feb.

Peterson, C.R. and Einstein, H.H. 1992. Manufacturing underground space, *Towards New Worlds in Tunnelling*, Balkema, Leiden.

Salazar, G.F., Kim, Y.W., Ioanou, P.G. Einstein, H.H. 1987. Computer based decision support systems in underground construction, *Proceedings Rapid Excavation and Tunneling Conference*.

Sinfield, J.V. and Einstein, H.H. 1996. Evaluation of tunneling technology using the "Decision Aids for Tunneling", *Tunnelling and Underground Space Technology*, Vol. 11, No. 4.

Xu, S., Grasso, P. Mahtab, M.A. and Einstein, H.H. 1998. Decision Aids in Tunneling, 10th Anniv. Issue, *World Tunnelling*.

Geotechnical Risks in Rock Tunnels – Campos e Matos, Ribeiro e Sousa, Kleberger & Lopes Pinto (eds) © 2006 Taylor & Francis Group, London, ISBN 0 415 40005 8

Fault zones and TBM

N. Barton
Nick Barton & Associates, Oslo, Norway

ABSTRACT: This paper is based on the lecture with this title that was given at the Aveiro Course on Geotechnical Risk in Tunnels in 2004. The present subject matter: *fault zones and TBM*, is richly illustrated with case record figures and pictures, mostly from the personal experiences of the author. The following main topics are covered:

1. Fault zone experiences in TBM tunnels in Italy, Greece, Kashmir, Hong Kong and Taiwan.
2. Fault zone cases in the Qtbm data base.
3. Attempts to be prepared using seismic and core logging and the Qtbm prognosis model.

1 INTRODUCTION

Three views of the world's first TBM tunnel in the UK Channel Tunnel Folkestone Warren area investigated by Beaumont in 1880. 1) A gravity-induced wedge fall-out in the chalk-marl. 2) The same tunnel with increased (cliff-induced) overburden showing stress-induced failure. 3) The same tunnel under the sea, with pore-pressure induced roof failures to bedding planes, with the added effect of **time**.

The failures shown in Figure 1 show classic rock mechanics features, which in the case of #3 are clearly enhanced by the effect of time. A rockmass classification of this tunnel, in relation to systematic *Q*-logging of the Channel Tunnel's main running tunnels driven 110 years later, was given by Barton and Warren, 1996.

The flat face of a large diameter TBM tunnel is not unlike a vertical rock slope. When a TBM cutter-head is withdrawn from a fault zone to (post) treat the rock mass, there may be a loosening effect such as illustrated from the slope excavation shown in Figure 2. The case of loosening in a fault zone shown in Figure 3 is from Grandori et al. 1995, from the Evinos-Mornos Tunnel in Greece.

2 THE CASE OF PONT VENTOUX, N. ITALY

A fault zone destroys much of the familiar tangential stress arch, and tunnel stability problems often arise as a result. High pressure inflow and falls of clay and rock blocks are other factors. The Italian Pont Ventoux HEP headrace tunnel was increasingly making a tangent to numerous faults, as shown in Figure 4.

The adverse effect on tangential stress (arching) is clear from Figure 4, but adverse water pressures were to prove the biggest problem with respect to the cutter-head getting stuck in these various fault zones at Pont Ventoux.

Another problem in fault zones is the grippers, and maybe also the shield i.e. delayed treatment of rock that actually requires pre-treatment. Figure 6, after Wanner, shows the implications in graphic form. Tangential stress indicators have been added in Barton, 2000.

What if the rock is too weak for the gripper action? Grippers usually have 'rib-spaces' (e.g. at 0.6 m spacing), to avoid crushing steel sets that are placed at this regular spacing. But if too many

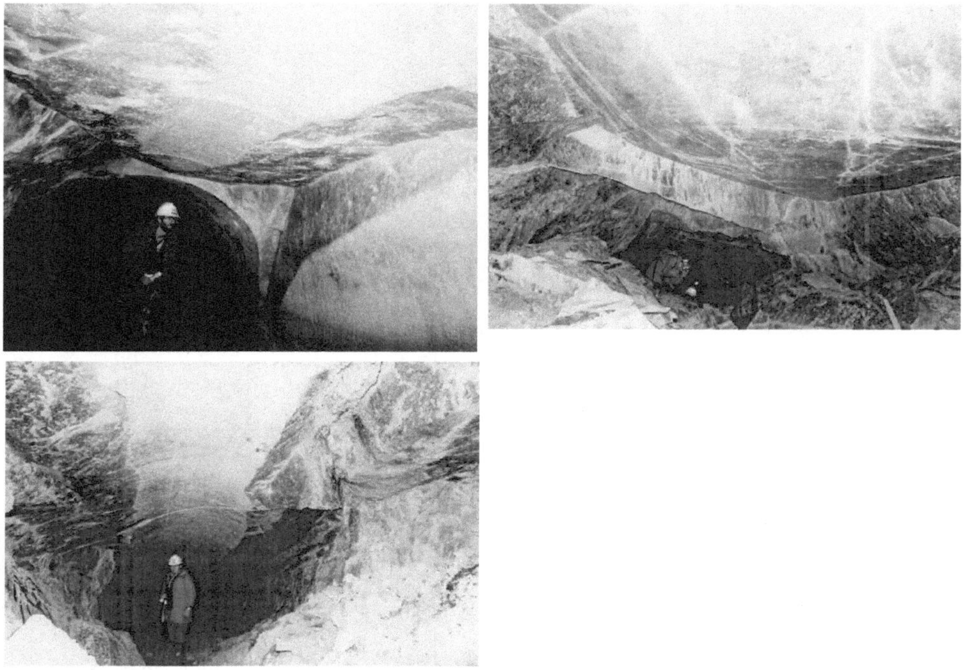

Figure 1. Three types of failure in the world's oldest TBM tunnel (1880).

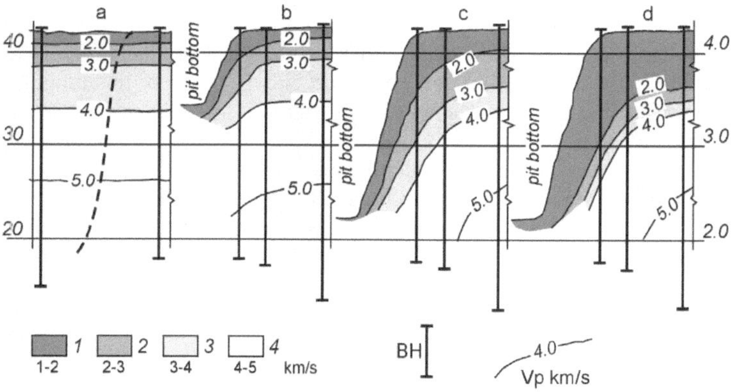

Figure 2. Effect of stress change and time on cross-hole seismic velocity at a Russian ship-lock. There is a 1 year delay between 'c' and 'd'. The reduced velocities are presumably caused by insufficient rock support where shear stresses are high, as also experienced in a 'flat' TBM tunnel face.

sets are needed (i.e. sets placed flange-against-flange) due to faulted rock, then there is the likelihood of crushing of the sets just where needed most. Such experiences from Pont Ventoux are shown in Figure 7.

Pont Ventoux HEP also had high (non-vertical) σ_1 stress, and very high water inflows, which were very adverse to stability in fault zones full of clay, silt, sand and crushed rock. Figure 8 illustrates this fatal combination where stability was good due to massive rock conditions.

The 7 km tunnel was parallel to a marked NW-SE trending valley, and also parallel to the foliation and to the (later discovered) fault zone swarms parallel to the valley side. The structural

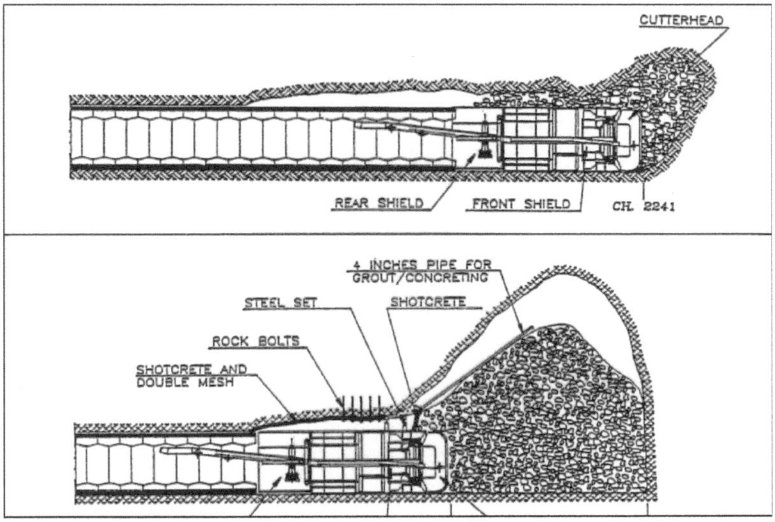

Figure 3. Loosening in a fault zone from Grandori et al. 1995, at the Evinos-Mornos Tunnel in Greece. We will return to this case later.

geology proved to be a disaster for the tunnel route, due to its near-parallel orientation to the later discovered faults.

Derailment of the train was frequent behind the back-up, due to build-up of a 'delta' of sand and silt washed out of the fault zone.

An NGI report written by the author in 1999 contained the following advice.

'Experience has shown that the extremely adverse conditions are caused by a combination of major water inflows, erosion of faulted material, and major void formation due to the generally adverse, sub-parallel fault orientations. The need to build a continuous steel liner (flange-to-flange) and further delays caused by cutter-head jamming, has given only 20 m advance in the last 7 months'.

'The outlook for future tunnelling is bleak if further members of the fault swarm lie sub-parallel, close to, or intersect the future tunnel. A drill-and-blast alternative of larger cross-section following the same route, or a revised route for continued TBM boring, or either tunnelling methods along a revised route, are three alternatives that need serious consideration'.

During 2004 the tunnel is expected to be completed by drill-and-blast from the other end of the tunnel, by-passing the rusting and abandoned TBM.

3 DEVELOPMENT OF THE Q_{TBM} METHOD OF PROGNOSIS

During the 6 months following the Pont Ventoux experiences, the Q_{TBM} method was developed (Barton 2000), based on an analysis of 140 TBM case records. The case records showed the following general 'deceleration' trends, when advance rate was plotted for various time periods. The classic equation AR = PR$\times$ U (where U = utilization), needs to be modified to the form AR = PR $\times$ T^{-m} to accommodate this fact.

The previously described Pont Ventoux fault zone performance of 7 months for only 20 m of advance, represents an average AR = 20/(7$\times$720) = 0.004 m/hr. This is almost off the bottom of the above chart, in the 'unpredicted events' area where various case record crosses (+) are plotted.

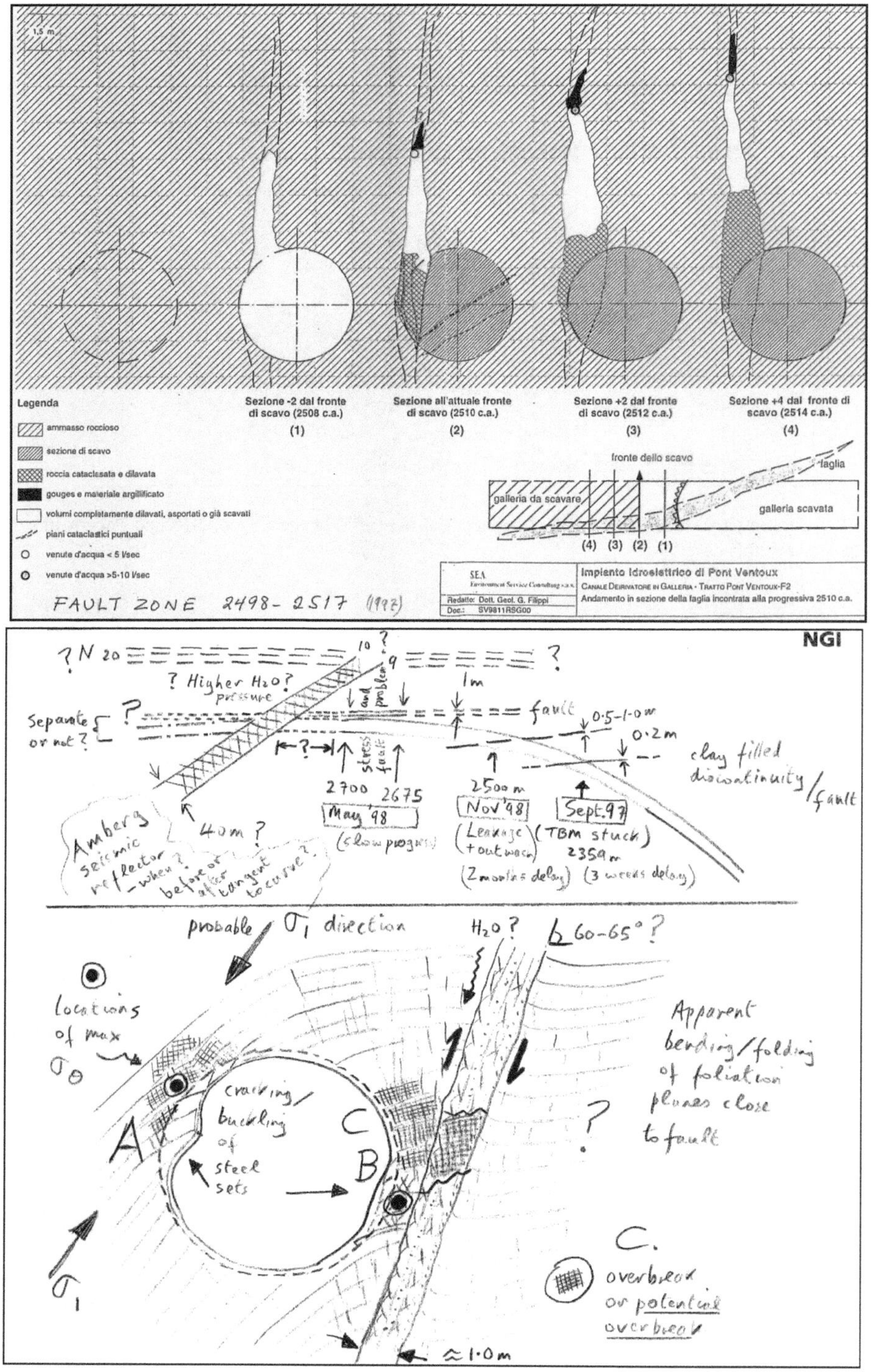

Figure 4. Two of the many fault-zone problems experienced at the Pont Ventoux HEP project in N. Italy (Barton 1999, NGI contract report).

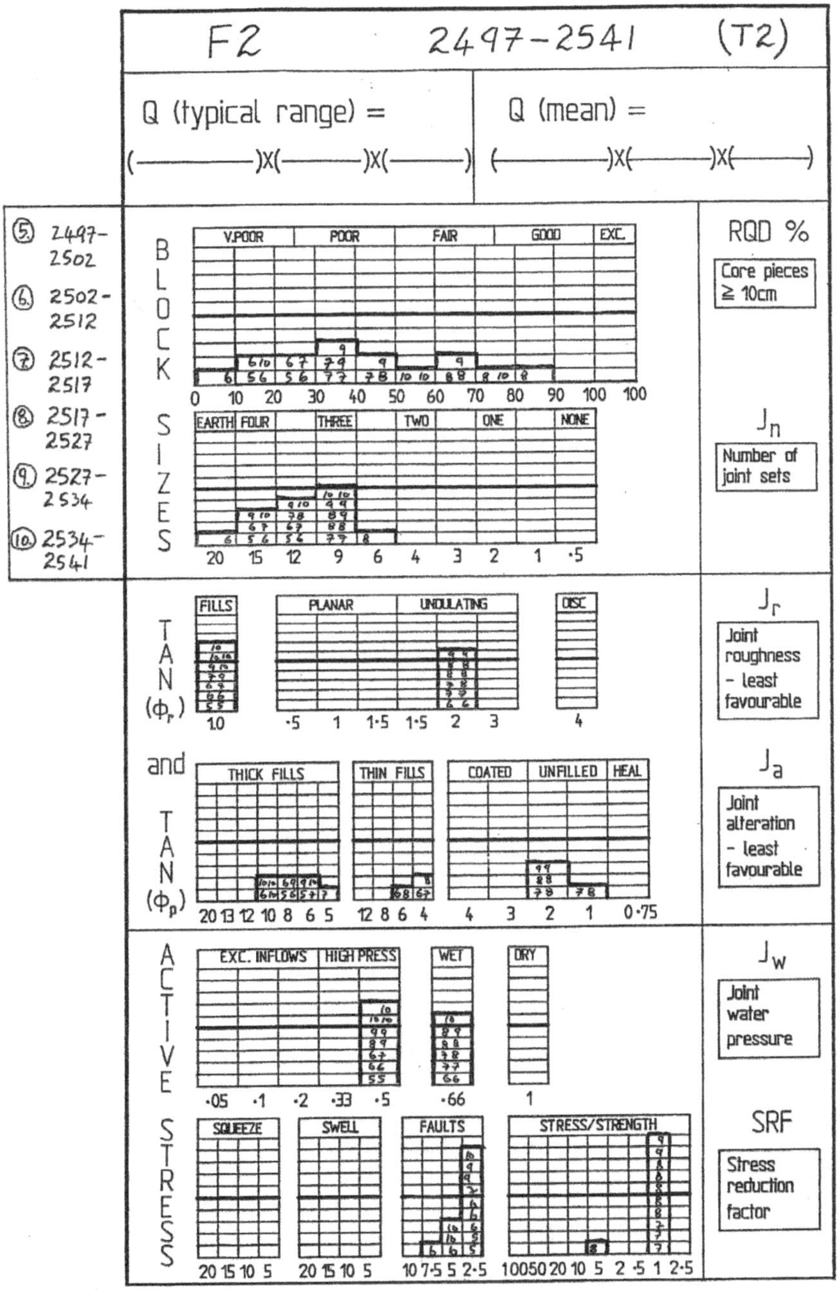

Figure 5. Q-histogram logging of part of an earlier section of the F2 fault at Pont Ventoux, from ch. 2500 m. Mixed RQD, high Jn, low Jr/Ja, low Ja, high SRF, and Q often < 0.1.

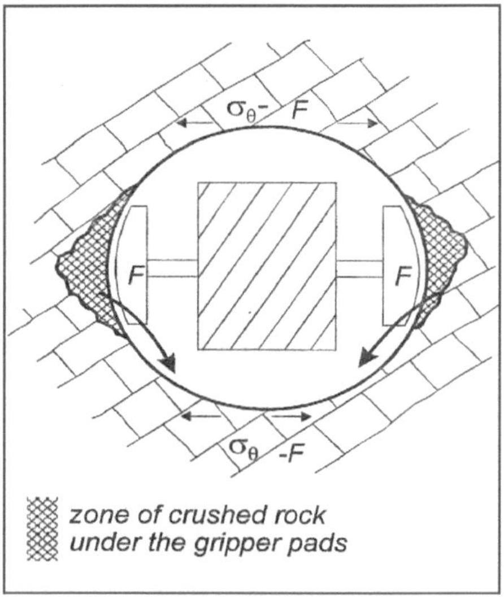

Figure 6. Adverse effect of gripper action where deformability is high, or alternatively; adverse effect of fall-out in haunches on gripper action. After Wanner, 1984, from Barton 2000.

Figure 7. Crushing of flange-to-flange steel sets in fault zone, by gripper action.

The gradients of *deceleration* (−m) given by the negative slopes of the TBM performance trend lines in Figure 12 are strongly related to *Q*-values when the quality is very poor (i.e. $Q < 1.0$) and so-called 'unexpected events' occur.

The 140 case records were analysed for 'best', 'average' and 'bad-ground' performance, as shown on the chart of log PR – log AR – log *T* given in Figure 14. 'Unexpected event' cases are shown as the steepest lines.

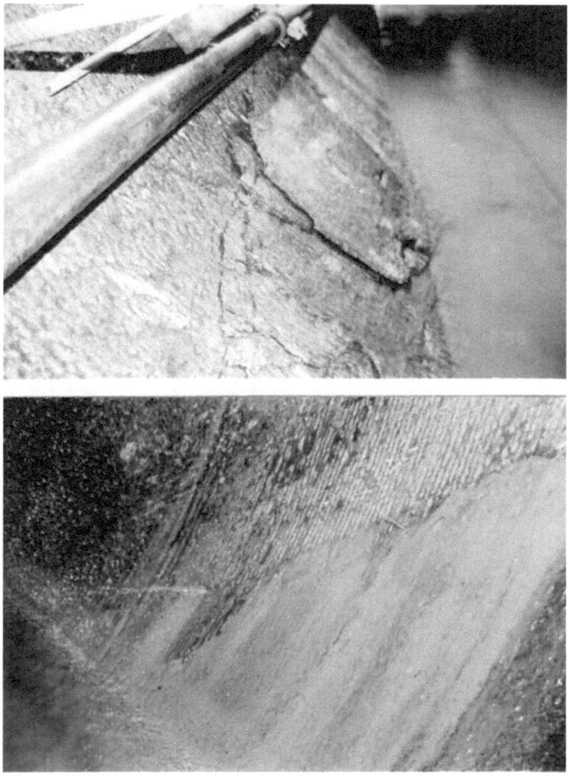

Figure 8. High anisotropic stresses and high water pressures seen where stability was good.

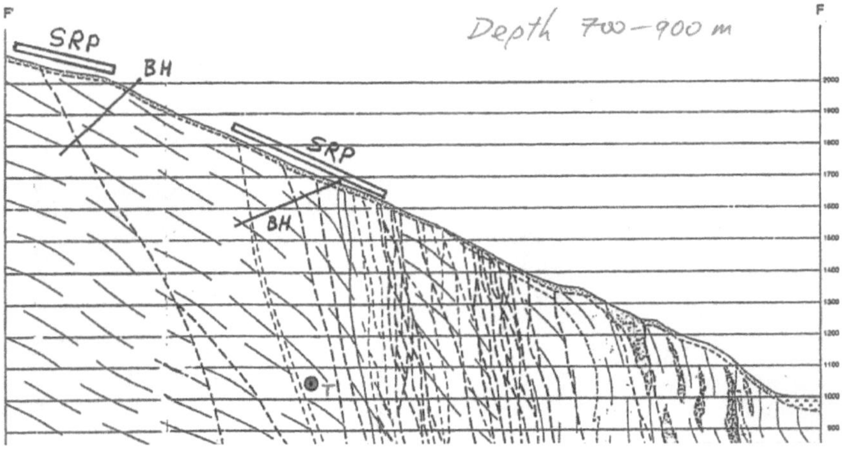

Figure 9. The tunnel was apparently 'too deep' for satisfactory geological investigations, judging by the 'missed' fault swarms shown here. In fact it was clearly not adequately investigated. BH boreholes, and SRP seismic refraction.

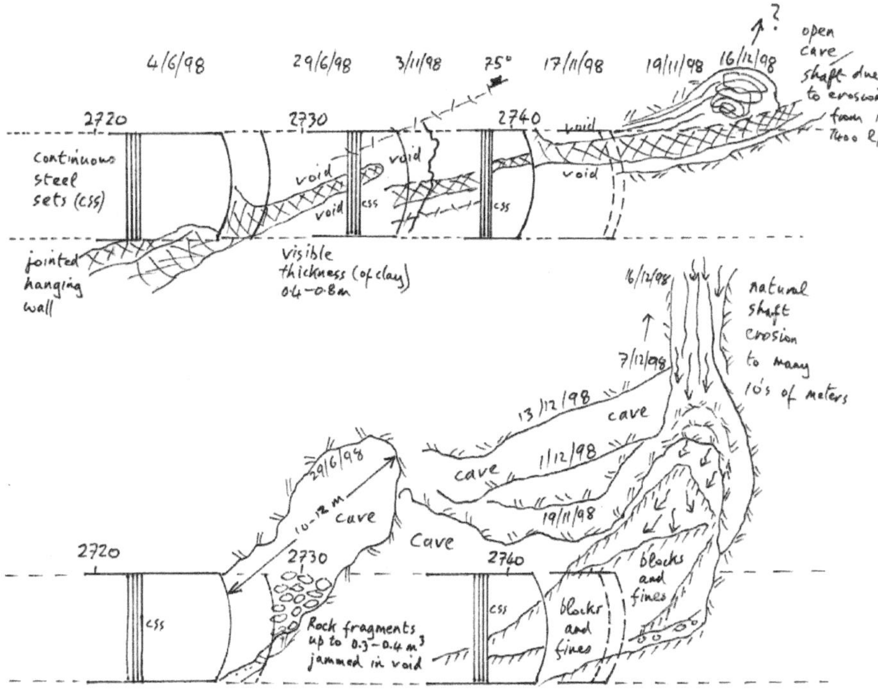

Figure 10. The Pont Ventoux TBM was stuck here for 6 months (due to blocked cutter head) from continuously falling blocks from the 'fault shaft', assisted by water and/or water pressure. These sketches are super-imposed on one sheet, from the geologist's daily logs (Barton 1999, NGI contract report).

Figure 11. View (120°) into inverted 'fault trench', and view of 'finger shield' and levels of water at the blocked face and behind the back-up (Barton 1999, NGI contract report).

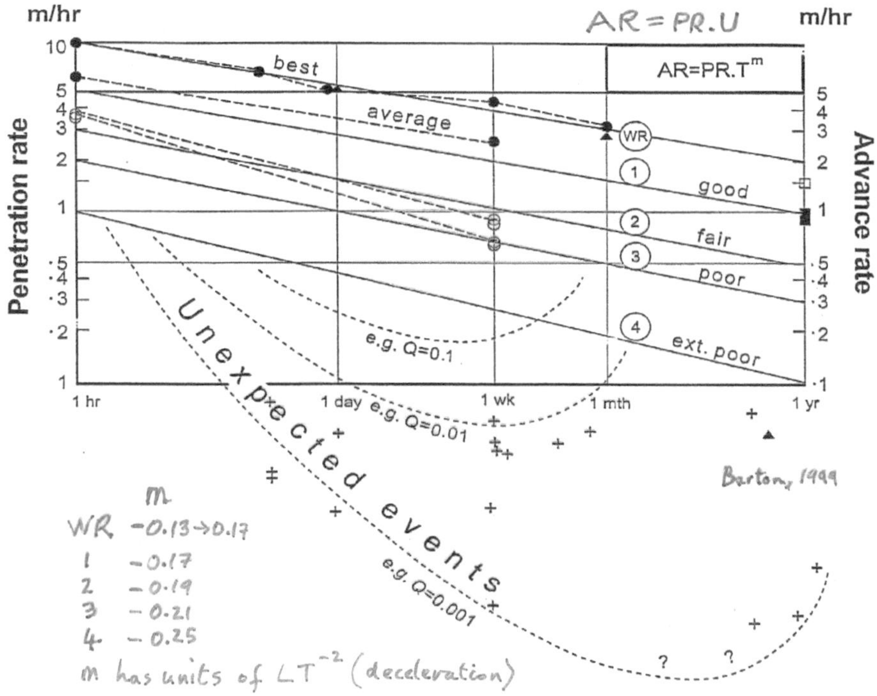

Figure 12. General trends of deceleration, and the adverse effect of 'unexpected events'. (Note PR = penetration rate, AR = actual advance rate, U = utilization = when boring, and T = time in hours) (Barton 2000).

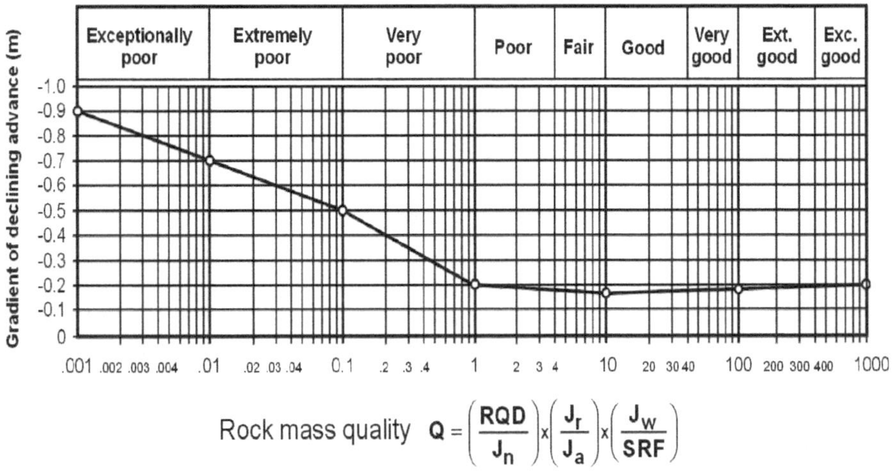

Figure 13. Preliminary estimation of deceleration gradient (−m) from the Q-value, is clearly of relevance for fault zones, as they have Q-values ≤0.1.

Using oilwell-stability sketches (from Bradley 1979), we can say that many faults (or boundaries of faults, where there is water), cause a 'ravelling' type of behaviour, like #4 in Figure 15.

The clay core of a fault (if present), may suffer squeeze, like #3. The well jointed case (#1) may be ideal for TBM due also to its favourable orientation, assuming little support is needed (giving

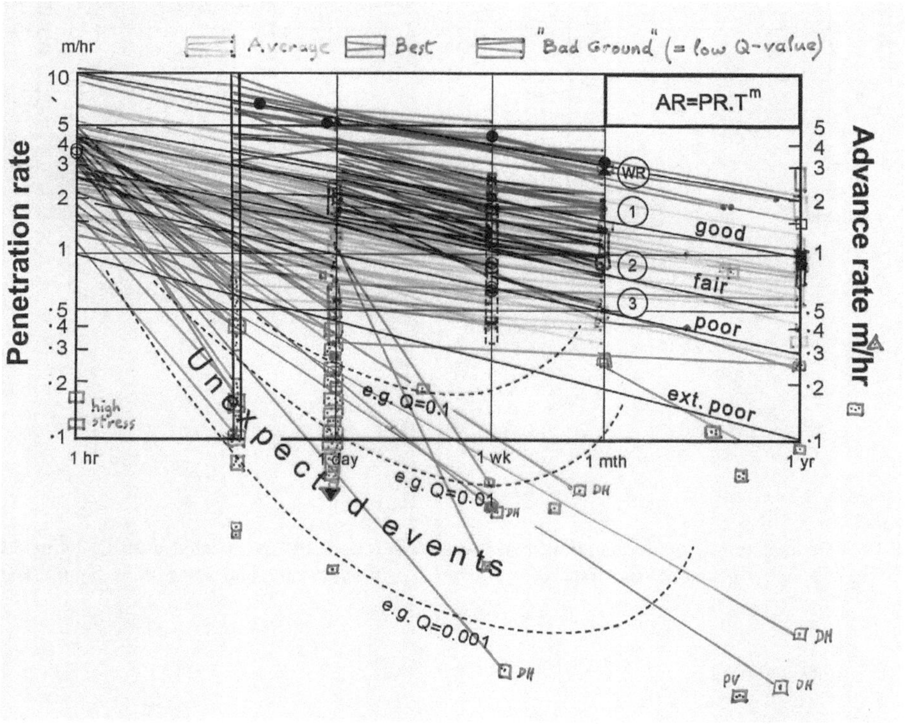

Figure 14. The 140 case records used to develop part of the Q_{TBM} prognosis model (Barton 2000).

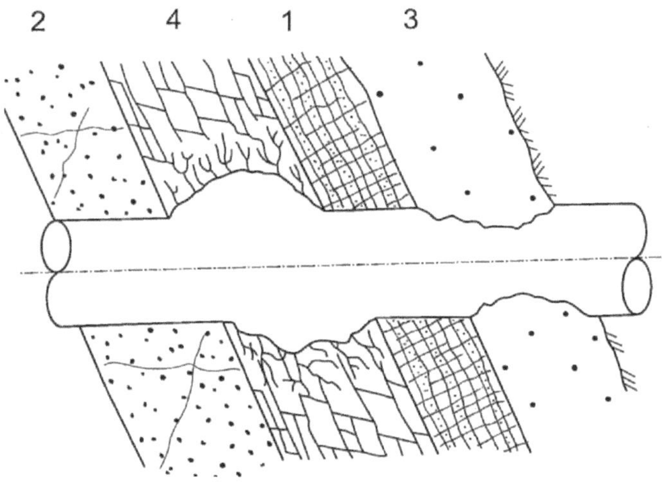

Figure 15. Four characters of ground from oil-well experience (Bradley 1979).

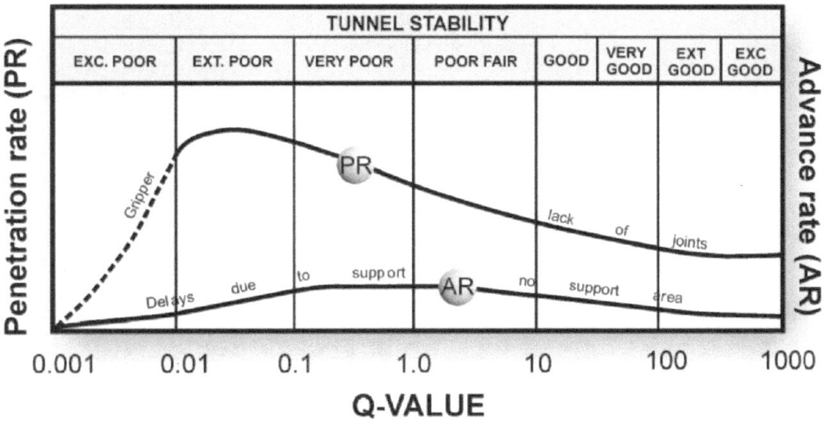

Figure 16. The Q-value descriptions of tunnel stability, which have general applicability, are however not sufficient on their own to describe TBM performance, due to the special cutter-rock and TBM – rockmass interactions (Barton and Abrahao 2003).

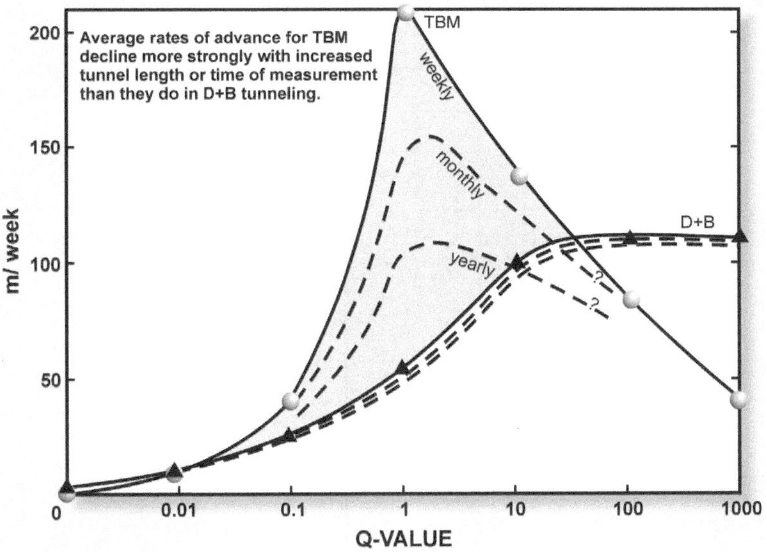

Figure 17. Comparing TBM and drill-and-blast in similar rockmasses, we see the need for a modified classification for TBM prognosis, and the effect of time is clearly 'adverse' for the TBM. The TBM is a supremely efficient method for 'central' rockmass qualities (Barton 2000).

both high PR and high AR). However, the sparsely jointed case (#2) may be tough to bore in hard abrasive rock (low PR, and consequently low AR).

4 DOUBLE-SHIELD TBM FOR MINIMISING MINOR GEOLOGICAL DELAYS

Use of double-shield TBM for PC-element building can solve minor stability problems without encountering significant delays, but when significant fault zones are intersected, the double-shield

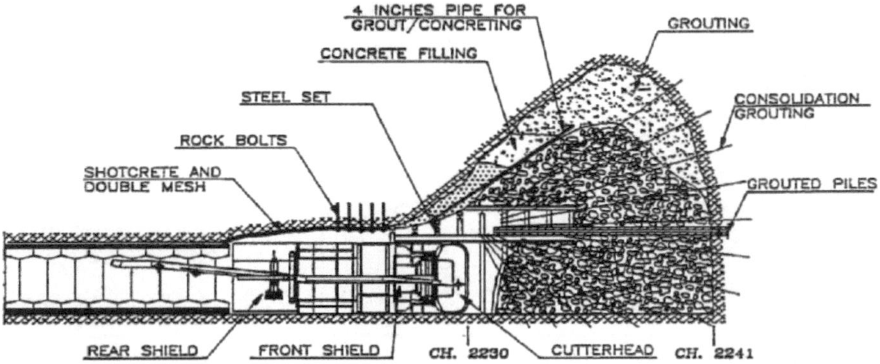

Figure 18. Detail from recovery operations described by Grandori et al. 1995. Despite the sophistication of double-shield operations (and their greater cost), hand-mining operations may be needed on occasion.

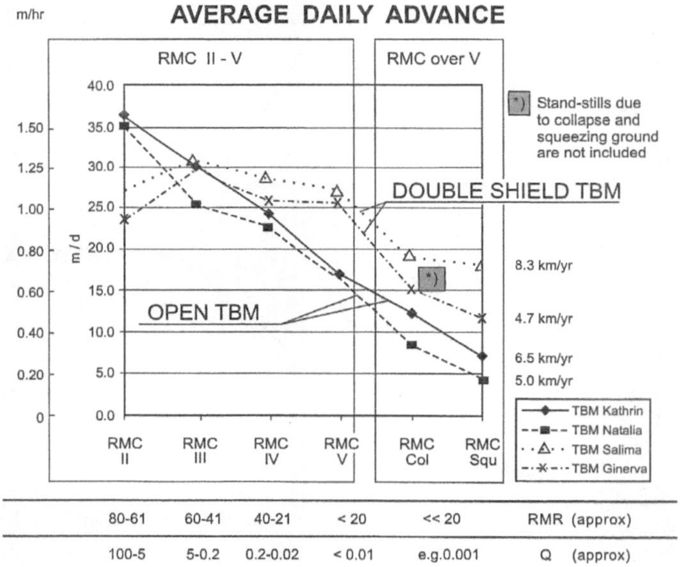

Figure 19. The Evinos-Mornos tunnel had two open TBM, and two double-shield TBM. Grandori et al. 1995 gave this comparative performance data. Tentative RMR and Q scales have been added by Barton, 2000. Note that 'stand-stills' (i.e. caused by fault zones) are not included.

may actually represent a hindrance to rapid recovery, as pre-treatment of the ground ahead is hindered by the long shields.

Despite the hard massive granites and gneisses at Guadarama, and the need for frequent cutter change, the overall efficiency of the 'continuous' thrust abilities described above, allows for a very shallow (excellent) gradient of deceleration (-m), as demonstrated by hand-written comments in the Figure 23 Q_{TBM} model chart shown.

Figure 20. Photograph of a modern double-shield machine, one of four used at the Guadarama tunnels in Spain (Photo: N. Barton).

Figure 21. PC-elements assembled under the protection of the tail-shield, allow 'continuous' operation of the TBM when dimensioned to take the thrust while re-setting grippers. Guadarama, Spain (Photo: N. Barton).

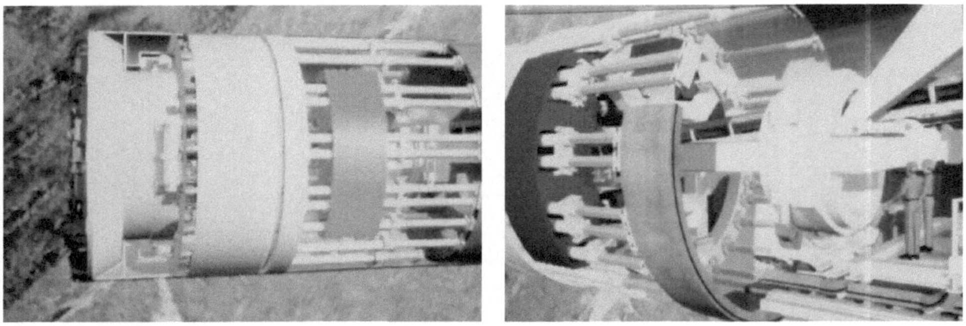

Figure 22. A computer animation of the 'continuity' of operations at Guadarama. Left shows thrust off PC-elements (with gripper re-set shown in green), right shows PC-element building during thrust off grippers (shown red) (Photo: N. Barton).

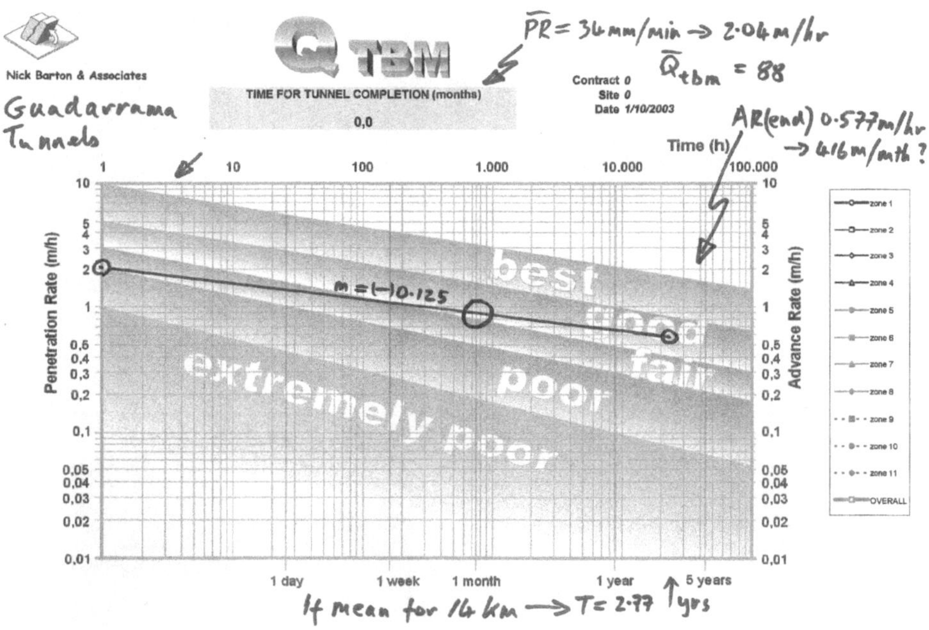

Figure 23. Despite frequent cutter changes, and a relatively low PR and therefore a high (unfavourable) Q_{TBM} value of 88 (see later), the high efficiency of 'continuous' boring at Guadarama gives an extremely small gradient of deceleration (−0.125), crossing 'normal' performance trends.

5 TBM PROBLEMS AT DUL HASTI HEP IN KASHMIR

We will now move further east into the foothills of the Himalayas, and record an extreme water and pebble/sand blow-out, plus stand-up time problems in inter-bedded phyllites, mixed with PR rates as low as 0.2 m/hr due to sections of massive abrasive quartzite.

The blow-out consisted of about 4000 m^3 of sand and quartzite pebbles (partly rounded by sub-terrainian flow) that buried the TBM, and an initial 60 m^3/min water inrush, that subsequently required a separate drainage tunnel to the valley-side. There was talk of miners having escaped from the flooded/buried TBM by escaping above the water, inside the air-ventilation ducts.

Figure 24. The price that one pays for this excavation efficiency is a huge materials fabrication and handling operation, at both portals if excavating from both ends.

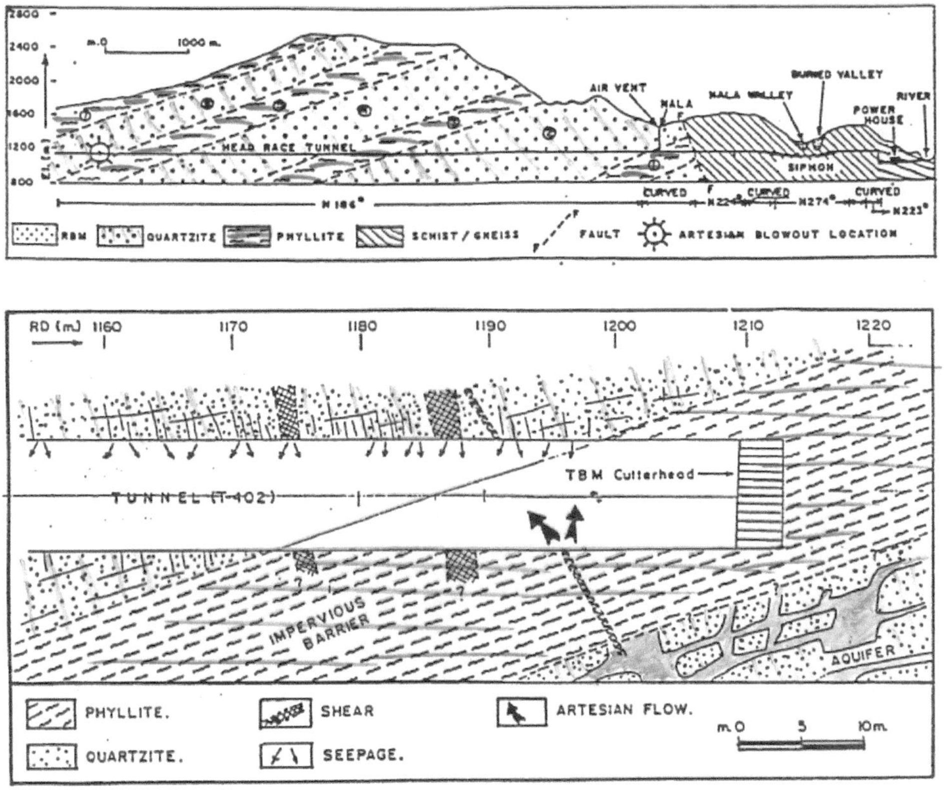

Figure 25. Simplified geology, and the location (ch. 1,215) of an extreme in-rush (or blow-out) of water, sand and rounded quartzite pebbles – at 750 m depth, which originated in the invert and was therefore not detectable by 'conventional' forward-and-upward probe drilling (which nevertheless was absent) (Deva et al. 1994).

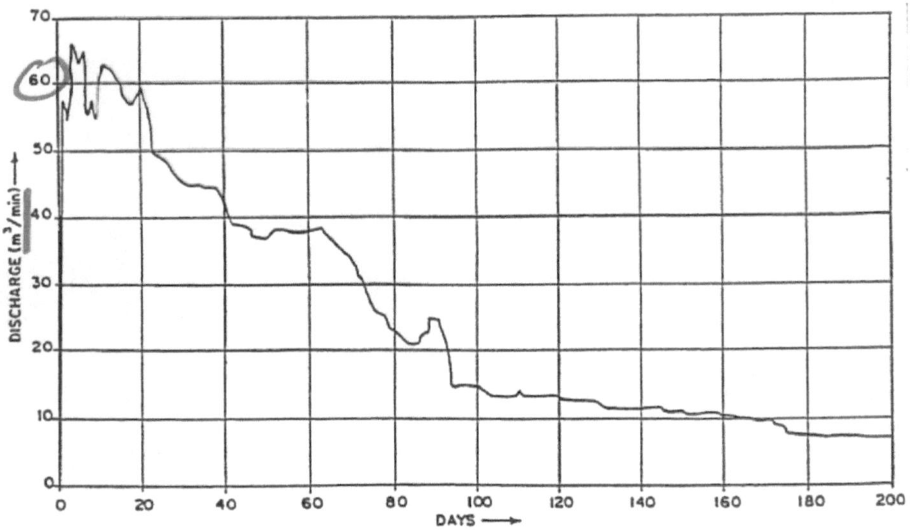

Figure 26. The initial 200 days of water in-inflow recordings that remained above 5 m^3/min even 5 years later. There was a 280 days delay due to this blow-out (Reva et al. 1994).

Figure 27. The location of the shear zone (left wall), and blow-out location (right wall/invert), as viewed some years after the event (Photo: N. Barton).

The fractured quartzite 'aquifer', sandwiched between impermeable phyllites, had its surface exposure more than a kilometer above and distant from the river valley. The connection of this 'aquifer' to the tunnel was by a minor shear, not even a fault.

The sheard, talcy phyllite was difficult to walk on, behaving like dry bars of soap. Blocks continued to fall from the sheared left wall, while the arch was in quite massive phyllites.

Figure 28. A subsequent location (around ch. 1630) of stand-up time problems in sheared, talcy phyllites, where a void of 3 to 4 m depth developed in the left wall, due to 'over-excavation' by the TBM, caused by the negligible stand-up time of this sheared rockmass (Photo: N. Barton).

Figure 29. A detailed view of the sheared phyllite next to the single shield TBM (Photo: N. Barton).

Figure 30. The TBM had been excavating more material than (πR^2) $\times$ *length* of advance, due to the stand-up time limitations of the sheared phyllites (Photo: N. Barton).

Figure 31. Block falls from the left wall continued during attempts to shore-up the void and back-fill with cement/sand sacks (Photo: N. Barton).

The Q-parameter logging provided an estimate of Q_{mean} as follows:

$$Q_{\text{mean}} = 0.07 \text{ in the sheared phyllite}$$
$$\text{Assume RMR} \approx 15 \log Q + 50 \qquad \text{(Barton 1995)}$$

From Figure 33:

$$1\,\text{m (without support)} \approx 1 \text{ hour stand-up},$$
$$5\,\text{m (no support until finger shield)} \approx 0.1 \text{ hr stand-up}$$

Borrowing the 'stand-up' time and 'roof span' data of Bieniawski 1989, we therefore find that the predicted stand-up time for an unsupported span (measured from last support) can have these adverse (much too short) magnitudes. These values easily explain why the TBM cutter-head was able to 'over-excavate'.

6 FAULT RELATED PROBLEMS AT SSDS TUNNEL F IN HONG KONG

The Tunnel F problems were mostly related with fault zones, and with the difficulty of pre-injection in a small-diameter TBM tunnel. This particular contract was completed by Skanska International, following re-negotiation of all the contracts after the withdrawal of the original contractor. The tunnel is in the top-left corner, going sub-sea, some 3 km from Tsing Yi Island to Stone Cutters Island, and underneath the world's second largest container port.

Unfortunately, the tunnelling contract consultants failed to detect and locate a major regional fault zone: the Tolo Channel fault zone, due in part to the difficulty of performing the sub-sea seismic profiling exactly as intended. Due to intense shipping activity close to the container port the seismic velocity profiles did not succeed in extending into the low velocity areas.

A summary of the situation at Tunnel F (when the author visited the tunnel for the first time was as follows):

- 481 m of the tunnel was completed in a previous project.
- 3098 m remained for the new contractor (Skanska).
- The Owner/Consultant expected 96 m/week, 204 m/week and 228 m/week (in poor, fair and good rock conditions – with a TBM inherited from the previous contractor).

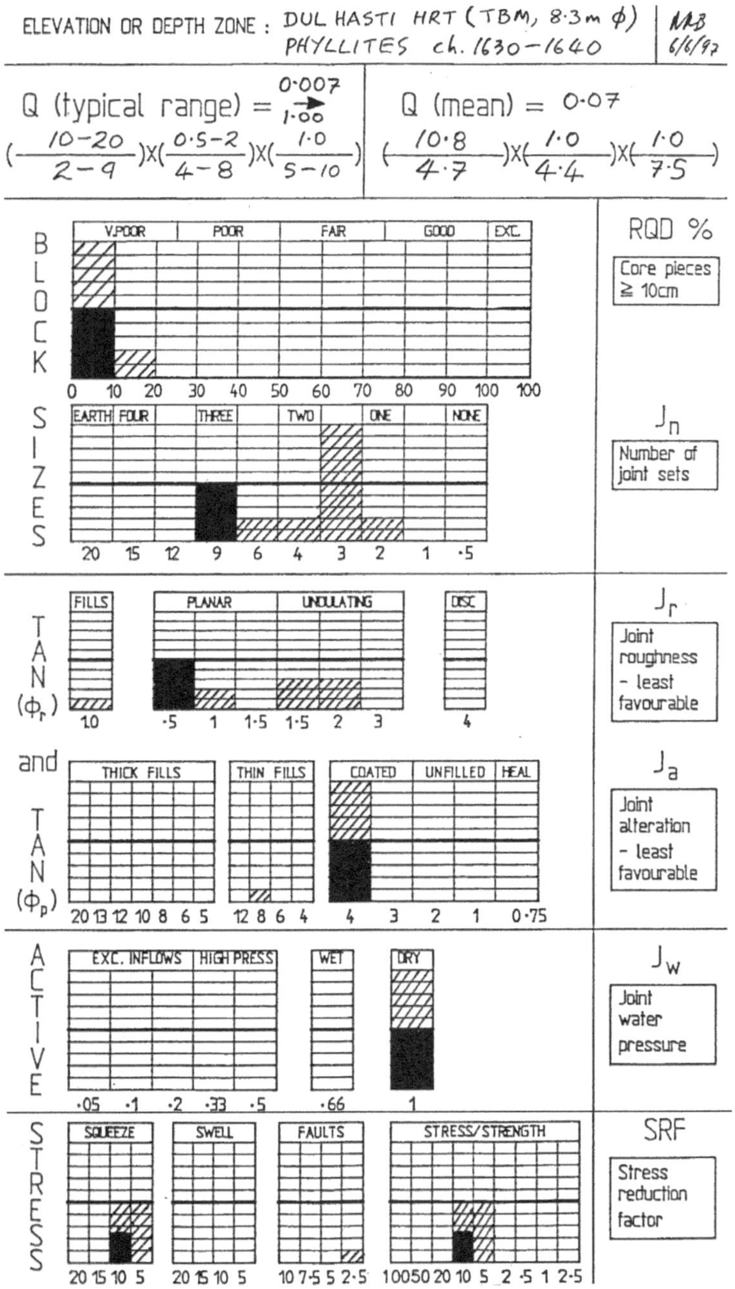

Figure 32. *Q*-parameter recordings in histogram format, for the sheared talcy phyllites illustrated previously. Note that due to contract re-negotiation, this TBM (abandoned by the original contractor) had progressed only 400 m in about 4 years (Barton 1997, NGI contract report).

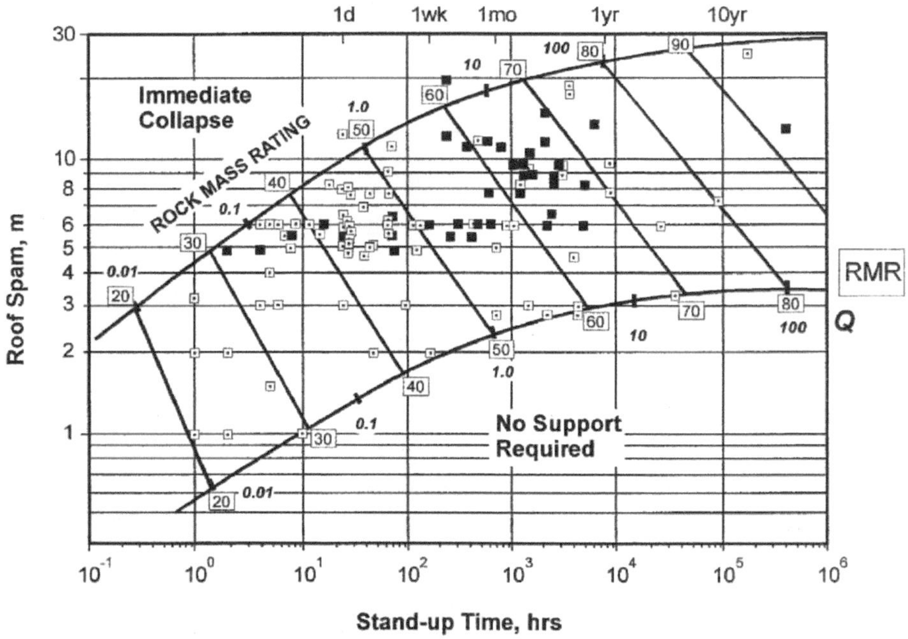

Figure 33. Stand-up time data plotted by Bieniawski, 1989, with RMR to Q conversion with the following equation from Barton, 1995.

- The conforming contract demanded 1 year for completion (of 3098 m).
- During 29 months, 2221 m of new tunnel was driven by Skanska.
- This represented an AR of 17 m/week (or AR = 0.1 m/hour) – which was 1/10 of the Owner/Consultant general expectation (and 1/3 of the conforming contract).
- Chainage 744–759 (15 m) had taken 8 months due to the need for hand-mining a by-pass round the stuck TBM in the first major fault zone (this represents a major 'unexpected event' with AR = 0.003 m/hour – and a mapped Q-value of about 0.001).
- Ch. 2622–2702 (80 m) took 4 months and 75 0000 kg of grout (average AR = 0.03 m/hr, i.e. also like an 'unexpected event', as plotted earlier).
- 887 m of tunnelling remained when the author started advising Skanska in 1999.
- There was a major regional (Tolo Channel) fault zone ahead, which had not been drilled or seismically profiled, due to heavy shipping traffic.
- Skanska decided to drill a long horizontal 'pilot hole' backwards from the shaft on Stonecutter's Island, to try to sample this major fault zone.
- The hole went only 731 m, as it was stopped by the Tolo Channel fault zone – despite three attempts at hole deviation.

7 ANALYSIS OF PILOT BOREHOLE (LH 01) CORE QUALITIES FOR INPUT TO Q_{TBM} MODEL

The 731 m of core recovered from the pilot hole sketched in Figure 41, provided Q-value input for much of the remaining tunnelling. The core was divided into five classes for convenience of description:

M = massive
S = slightly jointed
J = jointed

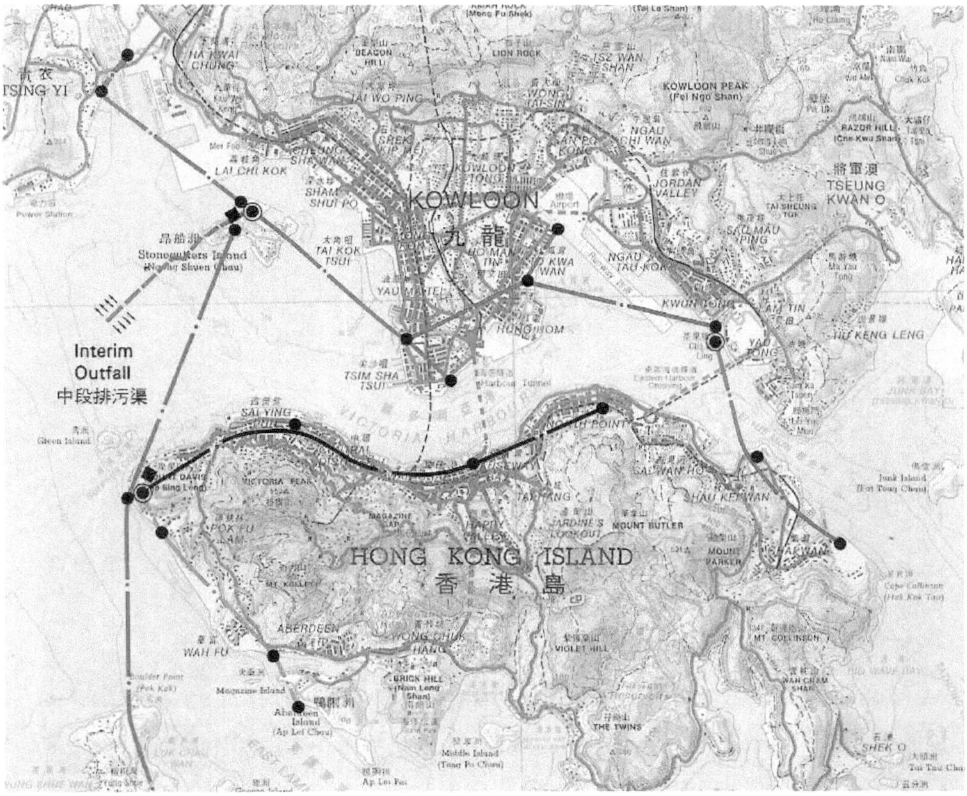

Figure 34. The Strategic Sewage Disposal System (initial TBM tunnels are nearer the Kowloon side of Hong Kong harbour). Much of the NE harbour area is now reclaimed land (SSDS Brochure).

Z = zone (weathered)
F = fault

Photographs showing these five classes are reproduced in Figure 43.

As shown in the next two figures, the availability of horizontal core data i.e. *parallel to the tunnelling direction*, is actually fundamental to a good TBM prognosis, especially when there is a marked anisotropy of structure.

While discussing horizontal core, it is appropriate to emphasise that if an efficient way of (multiple-hole) probe drilling and real-time analysis could be devised, for continuous application ahead of TBM, then operators could be better prepared for the changes that sometimes constitute serious 'unexpected events'.

8 USE OF Q_{TBM} PROGNOSIS AT TUNNEL F BASED ON CORE ANALYSIS

Three scenarios were modelled with the core data obtained from *Q*-logging of the horizontal core:

- First, with no pre-grouting improvement.
- Secondly, with grout-improved rock mass.... Net boring time.
- Thirdly, with the pre-grouting cycle time (approx.) included

Figure 35. The present intensely developed urban area of Kowloon and Hong Kong (with rare untreated exposed rock now preserved in places as 'architecture'), contrasts greatly with the geologist's possibilities of surface mapping in previous centuries. Today, smaller-scale features of the rock mass can be (temporarily) seen in cuttings, while large scale features of the terrain are largely obscured (Photo: N. Barton).

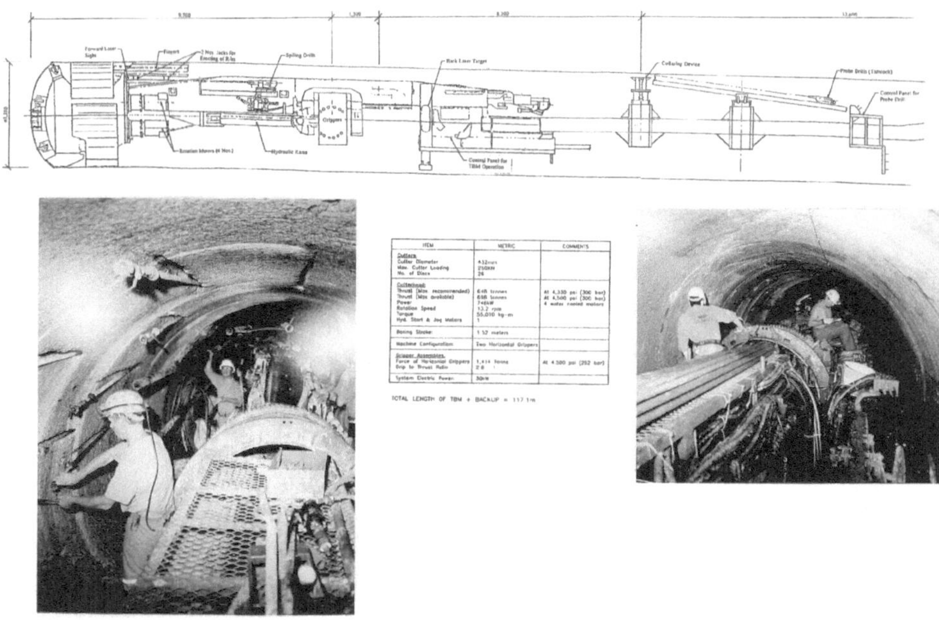

Figure 36. The confined conditions for performing pre-injection (Skanska photos).

Figure 37. With inflows as high as 1 m^3/sec locally, and huge quantities of grout, and several serious stability problems, an important aspect was worker morale – and drying of clothes between shifts (Photo: N. Barton).

Improvements in the rockmass properties caused by pre-grouting were based on the following types of arguments for particular rock classes: following Barton 2002.

RQD increases e.g. 30 to 50%
J_n reduces e.g. 9 to 6
J_r increases e.g. 1 to 2 (due to sealing of most of set #1)
J_a reduces e.g. 2 to 1 (due to sealing of most of set #1)
J_w increases e.g. 0.5 to 1
SRF unchanged e.g. 1.0 to 1.0

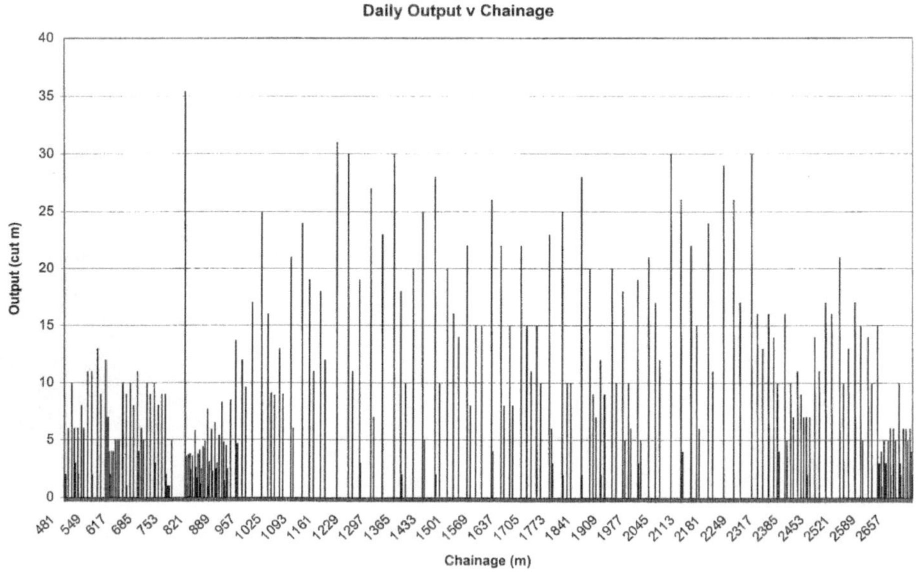

Figure 38. Daily advance versus chainage, from beginning of Skanska project (ch.481) to time of consultant's visit. The 'dark' areas are fault zones, or where particular water control problems were experienced (Skanska data) (Nick Barton & Assoc. 2001, contract report).

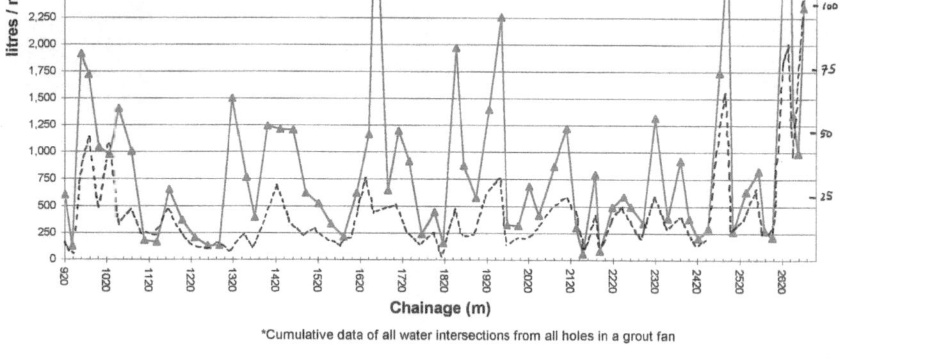

Figure 39. Super-imposed water inflows (cumulative, all holes) and OPC+MFC grout take (dotted lines), show general correlation (Skanska data) (Nick Barton & Assoc. 2001, contract report).

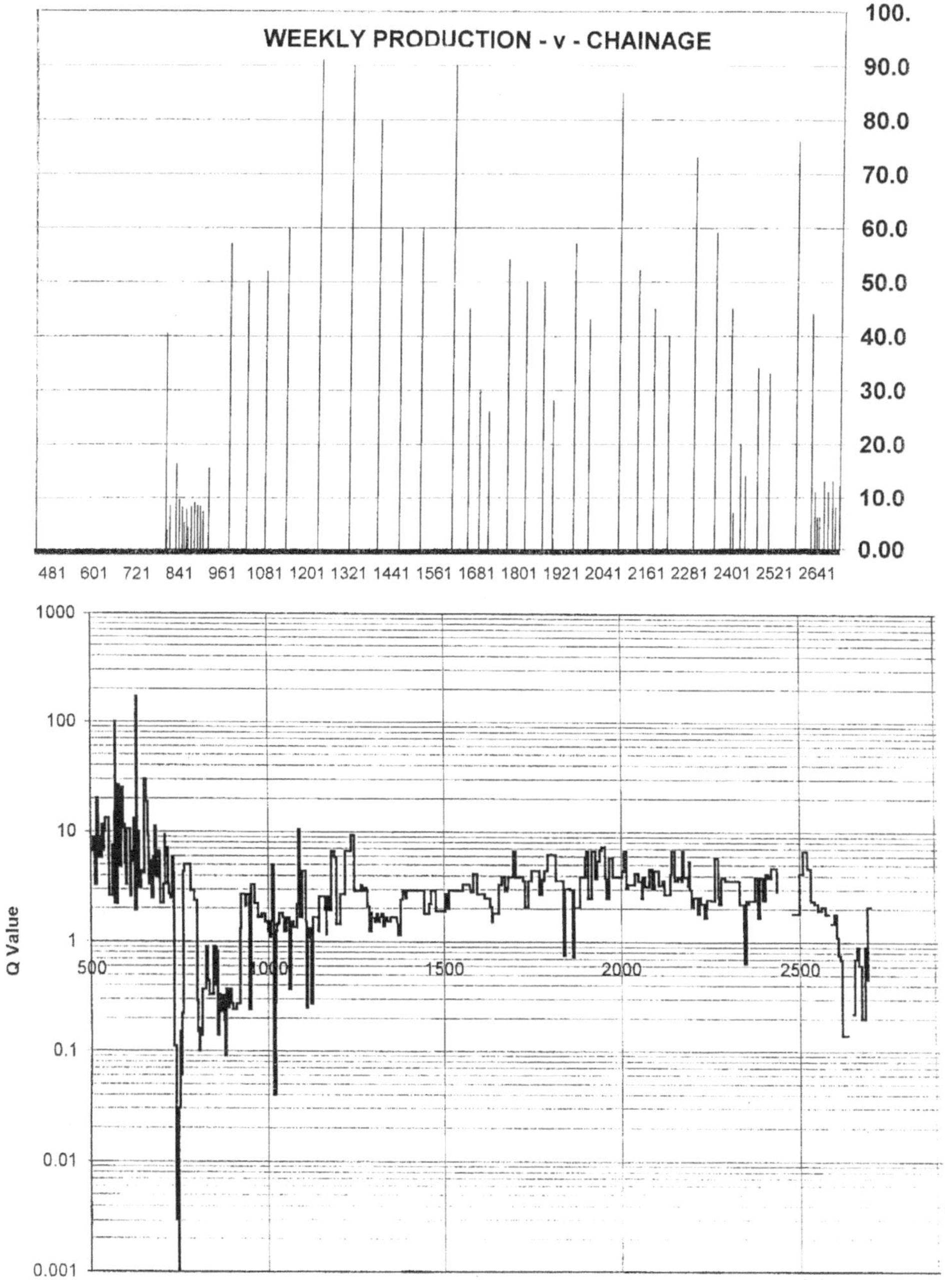

Figure 40. Correlation of logged Q-values and weekly advance, which was below 10 m/week on numerous occasions due to faulting and water control delays (Skanska data) (Nick Barton & Assoc. 2001, contract report).

Figure 41. The remaining 887 m of TBM-driven tunnel included the unexplored Tolo Channel fault zone. Only a few meters of this could be cored from the shaft location, yet the TBM later managed to penetrate the fault zone due to the pre-grouting effect on the rock mass (NB sketch of situation in 1999.) (Nick Barton & Assoc. 2001, contract report).

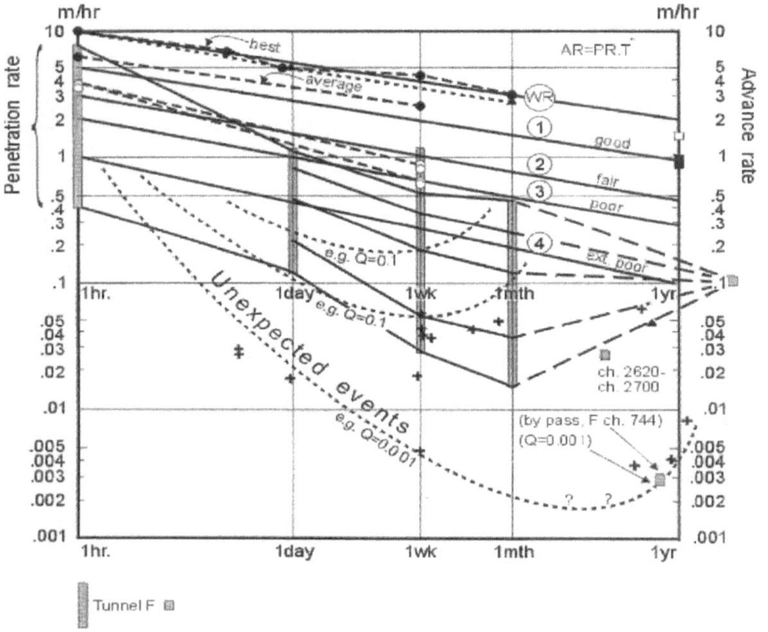

Figure 42. A plot of the recorded PR and AR performances of Tunnel F, superimposed on the Q_{TBM} chart of log PR – log AR – log T (Nick Barton & Assoc. 2001, contract report).

Before pre-grouting	$Q = 30/9 \times 1/2 \times 0.5/1 = 0.8$
After pre-grouting	$Q = 50/6 \times 2/1 \times 1/1 = 17$

With similar improvements in the different rock classes, due to appropriate assumptions, following recommendations in Barton, 2002, there is a reasonable expectation of improving rock mass properties through the pre-grouting that was an almost standard procedure ahead of the **TBM**.

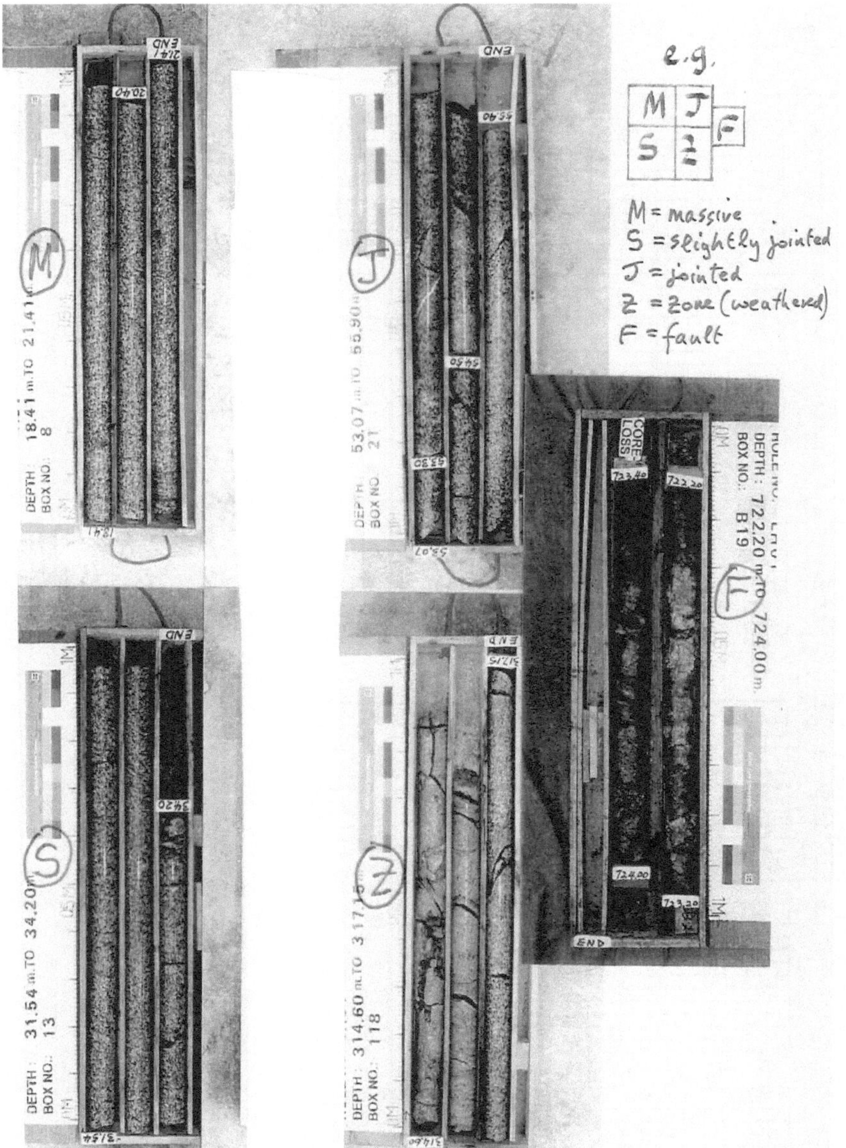

Figure 43. Photographs of the five selected rock classes which, when Q-parameter logged, gave the approximate statistical frequencies of these five classes (Nick Barton & Assoc. 2001, contract report).

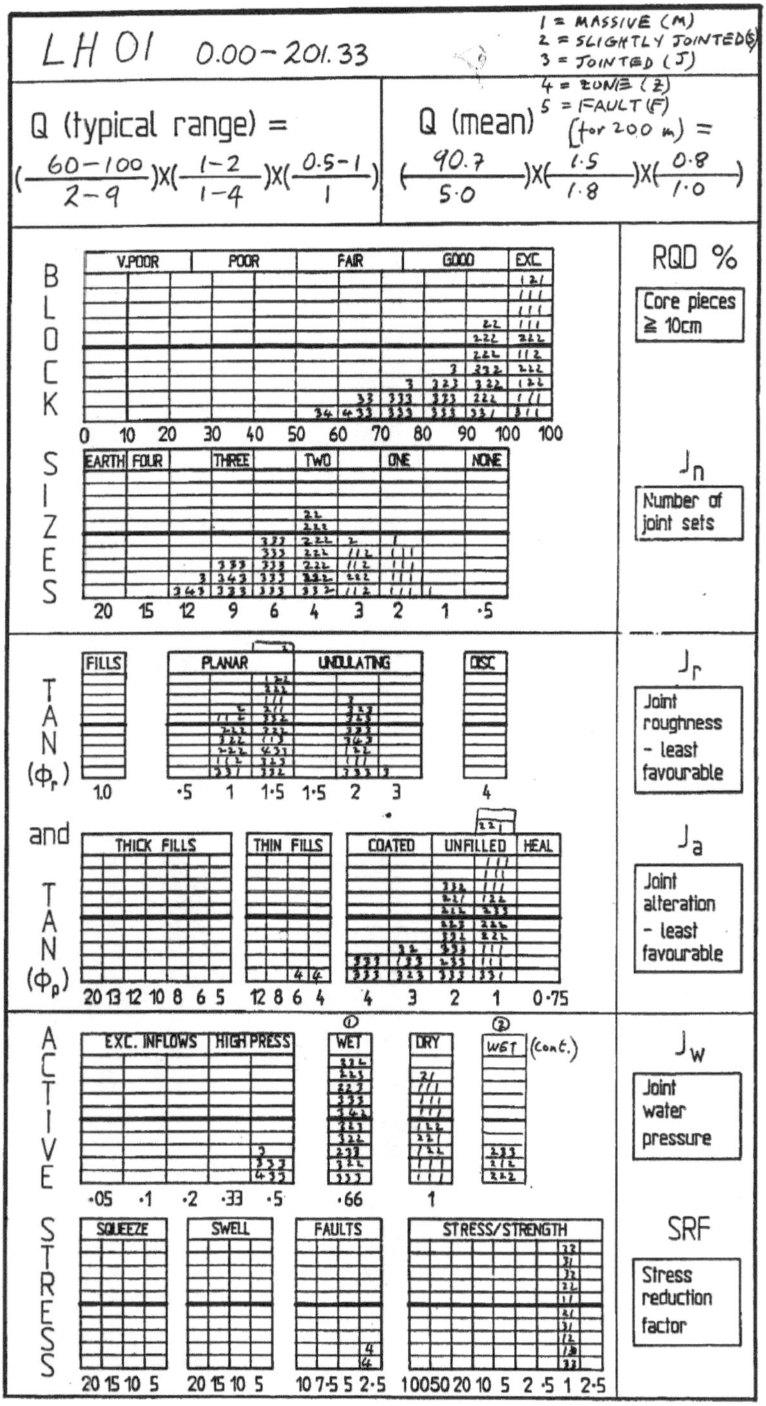

Figure 44. Q-parameter histogram logging of frequency of occurrence and ratings for the five rock classes, in the first 200 m of hole LH 01 (i.e. the last 200 m of the tunnel) (Note numbers 1–5 in each histogram, corresponding to rock class.) (Nick Barton & Assoc. 2001, contract report).

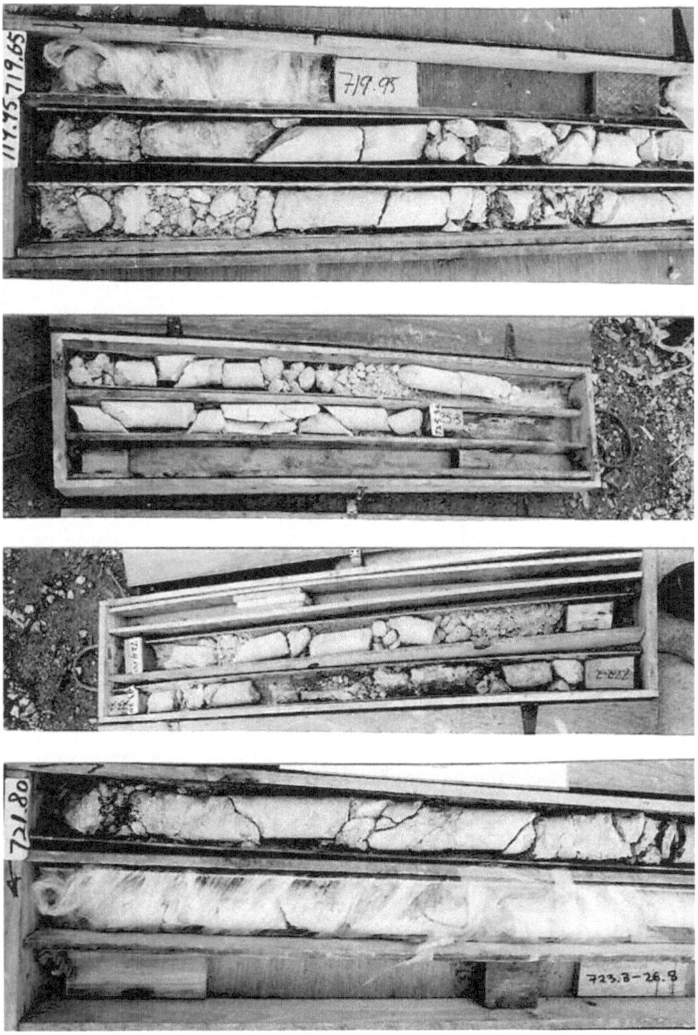

Figure 45. Three attempts were made to (deviate) drill into the Tolo Channel Fault Zone. These are the results – from the end of the 720 m long horizontal pilot hole (Nick Barton & Assoc. 2001, contract report).

Table 1. Example of rock mass and tunneling improvements that might be achieved by pre-injection. In poorer quality rock masses there could be greater improvements, in better quality rock masses it may be unnecessary to pre-grout (Barton 2002).

Before pre-grouting	After pre-grouting
$Q = 0.8$ (very poor)	$Q = 16.7$ (good)
$Q_c = 0.4$	$Q_c = 8.3$
$V_p = 3.1$ km/s	$V_p = 4.4$ km/s
$E_{mass} = 7$ GPa	$E_{mass} = 20$ GPa
B 1.6 m c/c	B 2.4 m c/c
S(fr) 10 cm	none

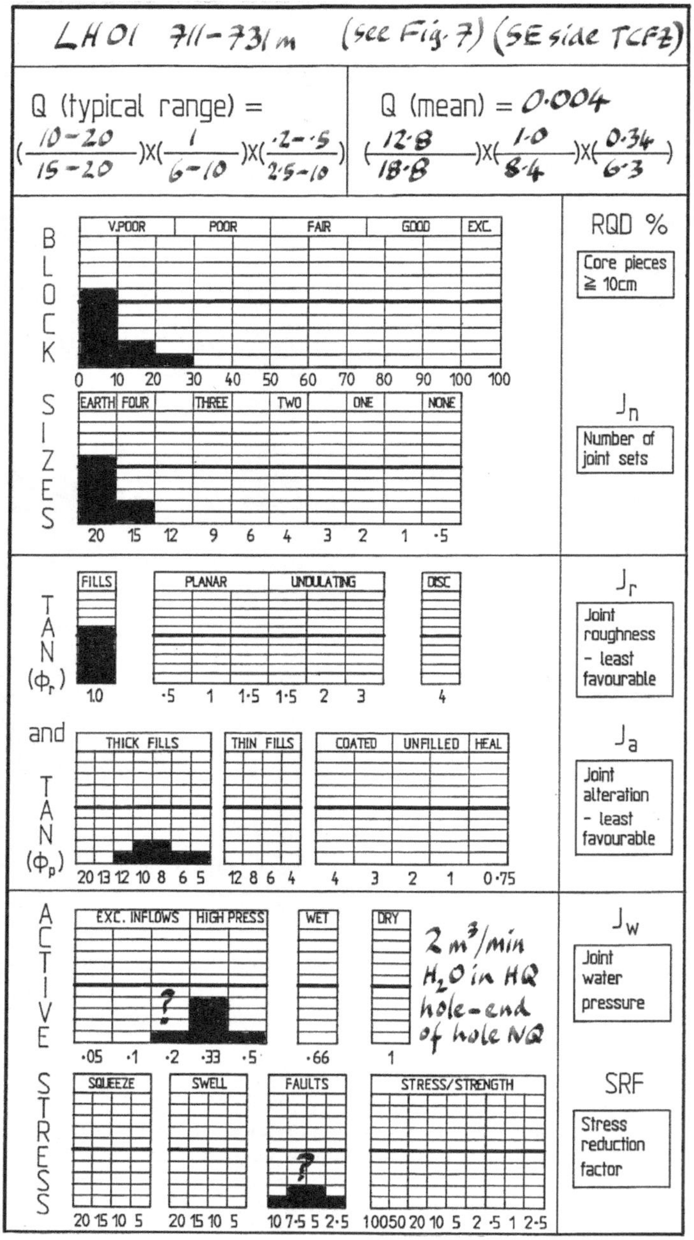

Figure 46. The Q-parameter characteristics of the fault zone (all that could be sampled) all plot 'to the left'. Q_{mean} = 0.004, i.e. needs improvement by pre-grouting (Nick Barton & Assoc. 2001, contract report).

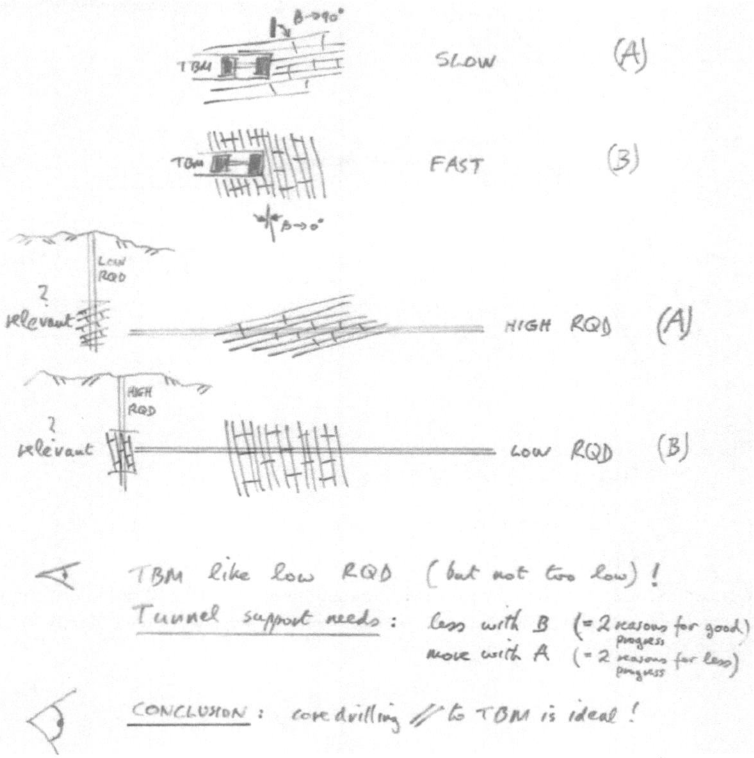

Figure 47. Sketches to demonstrate that horizontal boreholes (and therefore analysis of horizontal core) is the correct way to gather information for TBM prognoses when anisotropy of structure (bedding, foliation, schistocity, dominant jointing) is marked (Barton 2002).

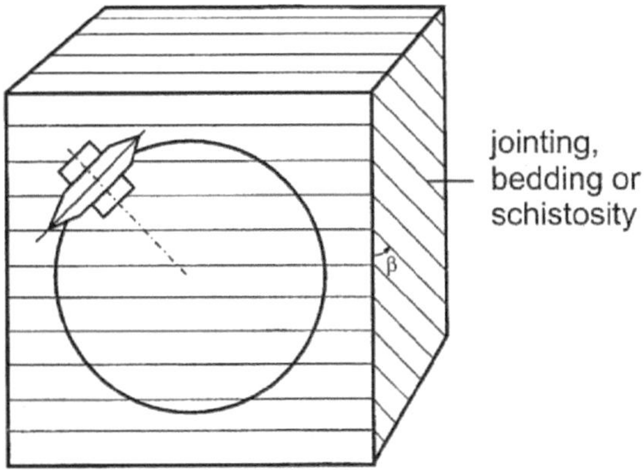

Figure 48. Ideal orientation of structure for best TBM penetration rate and advance rate (due to minimised support needs) is when angle β is small. Worst is when it is closer to 90°, as more rock has to be broken, without so much assistance from the existing structure (Barton 2002).

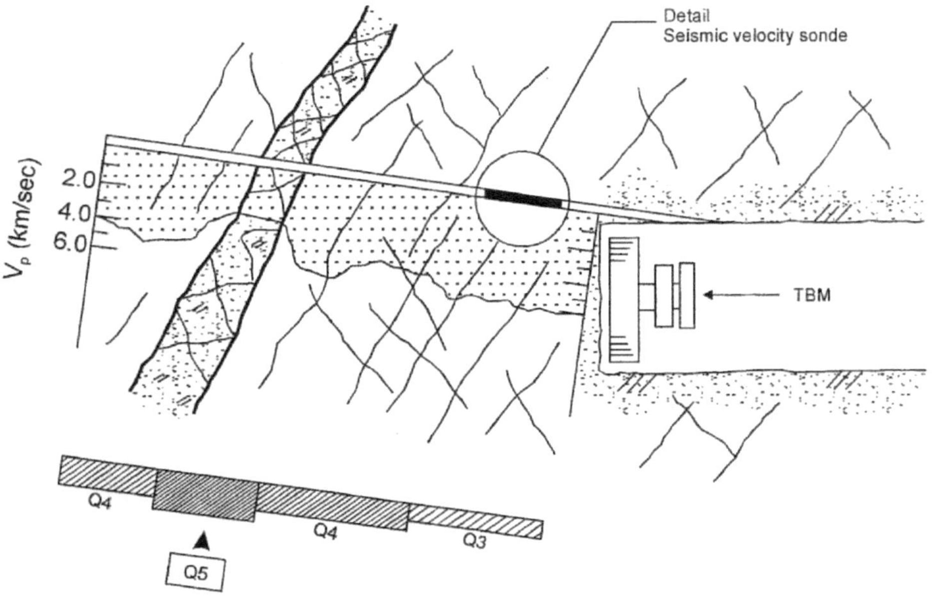

Figure 49. Conceptual use of probe holes for seismic logging ahead of TBM, and derivation of real-time information on e.g. the approaching Q-values, and perhaps need for pre-injection – not only for water control, but also for improving rockmass conditions.

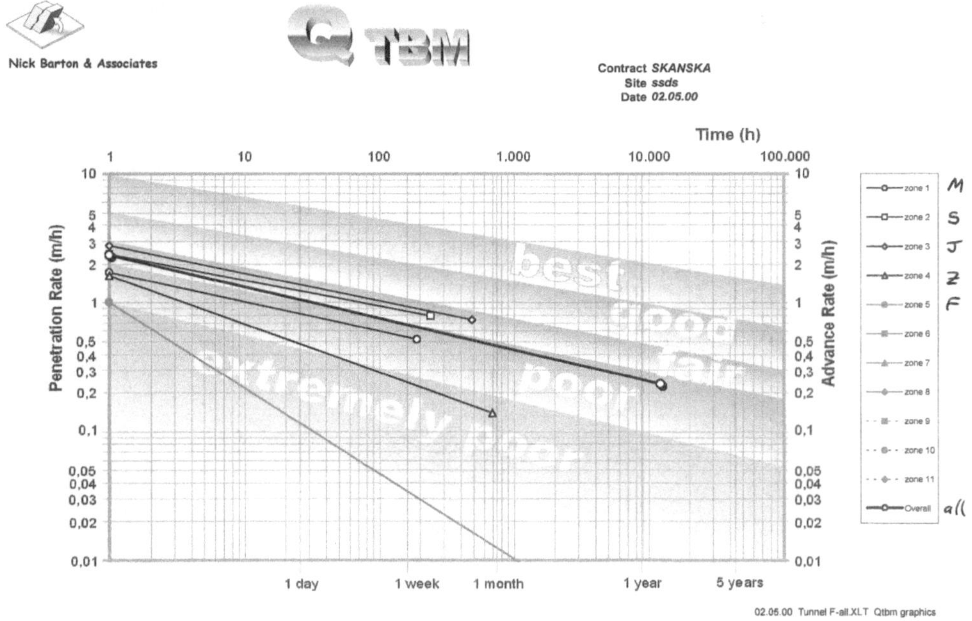

Figure 50. No pre-grouting improvements.... more than 1 year needed.

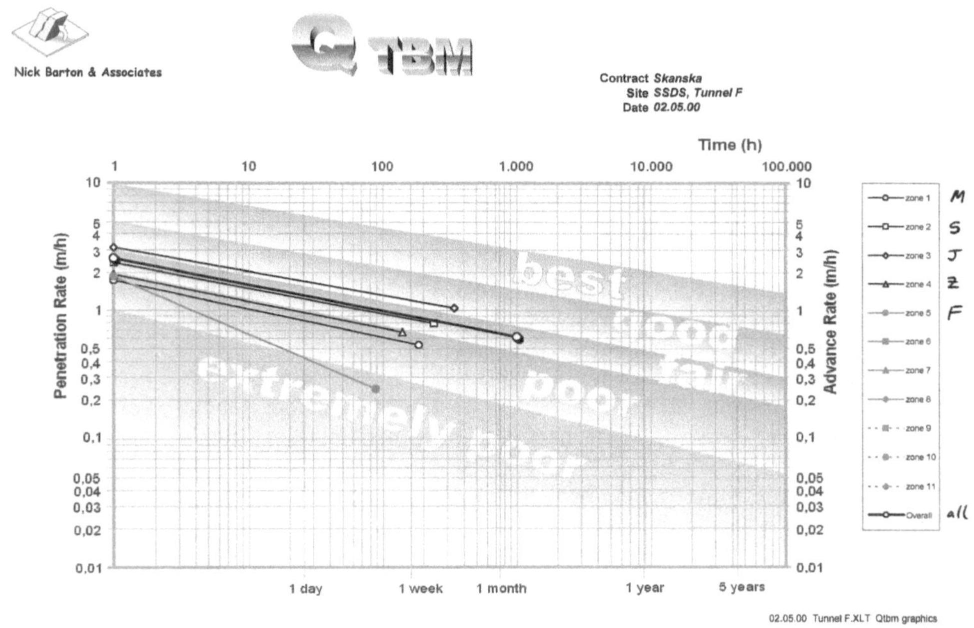

Figure 51. Net boring time with pre-grouting improvement. Note reduced gradients.

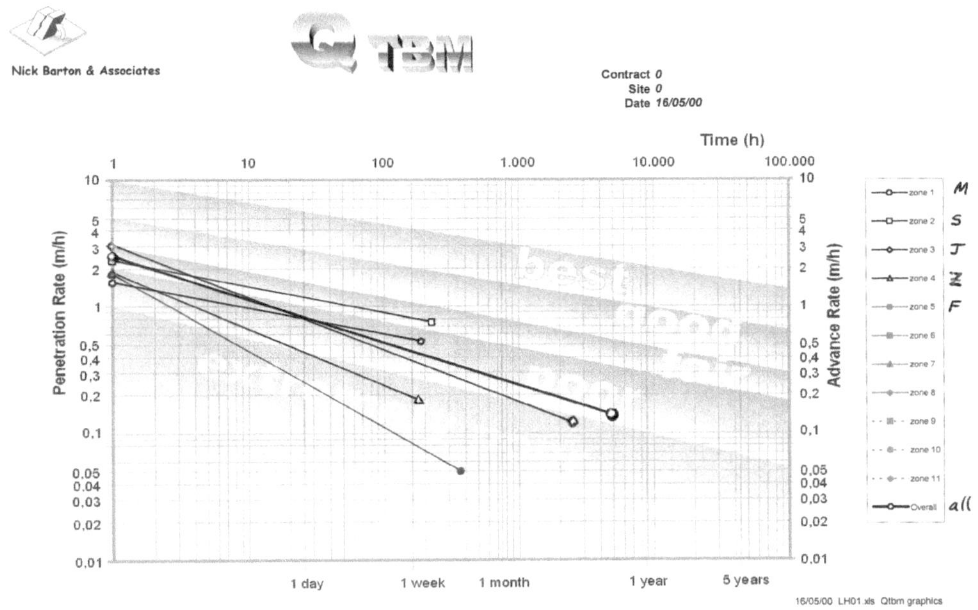

Figure 52. Total tunnelling time with pre-grout cycles (two estimates). Actual time was 2 months for the last 700 m. (Total time for Skanska's 3097 m was 37 months.)

The following is a summary of the results obtained with the Q_{TBM} modelling predictions:

ZONES M, S, J, Z (massive, slightly jointed, jointed, weathered zone)
$L = 710$ m, 19 weeks **with** pre-injection (needed for water control)
$L = 710$ m, 6.2 weeks **without** pre-injection

ZONE F (fault zone)

$L = 21$ m, 2.3 weeks **with** (successful) pre-injection
$L = 21$ m, **1.4 years without** pre-injection

The author's prediction of 3097 m remaining tunnel (following 2270 m known) was a total of 3.5 years. Actual completion was 37 months or 3.1 years. The pre-injection performed by Skaska had perhaps an even more beneficial result than assumed.

Why do fault zones take so long with TBM?

Practical reasons are illustrated in Figures 3, 10 and 54 (below, from Robbins 1982).

Theoretical (or *theo-empirical*) reasons are given in the following simple equations, from Barton, 2000.

There are unfortunately very good 'theo-empirical' reasons why fault zones are so difficult for TBM (with or without double-shields).

We need three basic equations to start with

1. $AR = PR \times U$ (all TBM must follow this)
2. $U = T^m$ (due to the reducing **u**tilization with time, **a**dvance **r**ate decelerates)
3. $T = L/AR$ (obviously time needed for length L must be equal to L/AR)

Therefore we have the following

4. $T = L / (PR \times T^m)$ (from #1, #2 and #3)
5. $T = (L/PR)^{(1/1+m)}$
6. This is very important equation for **TBM**, if one accepts that (−)***m*** is strongly related to Q-values in FAULT ZONES (as shown in the repeated figure below)
7. It is important because very *negative* (−)m values make $(1/1+m)$ *too big*
8. If the fault zone is wide (large L) and PR is low (due to collapses etc.) then L/PR gets too big to tolerate a big component $(1/1+m)$ in equation 5.
9. It is easy (too easy) to calculate an almost 'infinite' time for a fault zone using this 'theo-empirical' equation. The writer knows of three permanently buried, or fault-destroyed **TBM** (Pont Ventoux, Dul Hasti, Pinglin). There are certainly many more, and the causes are probably related to equation 5 logic.

The author's prediction for 3097 m remaining tunnel (following 2270 m known) was a total of 3.5 years. Actual completion was 37 months or 3.1 years. The pre-injection performed by Skanska had perhaps an even more beneficial result than assumed.

Fault zones will remain a serious threat to TBM tunnelling as we now know it, unless the extremely poor rock mass qualities associated with fault zones *can be improved by pre-grouting*.

9 BENEFIT OF PRE-INJECTION IS GREATEST WHEN $Q < 1.0$

In drill-and-blast tunnelling, thorough pre-injection around the complete (360°) profile can be more easily performed than when '*there is a TBM in the way*'. The 20–24 hours (approx.) needed to drill and inject 20–40 holes of about 25 m length is balanced by relatively trouble-free advance (of e.g. 4 or 5 rounds) through this (previously) bad ground, until the next cycle of pre-injection is performed to secure the next rounds. Typically 20–25 m/week tunnel advance can be achieved on a regular basis, despite the (previously) bad ground.

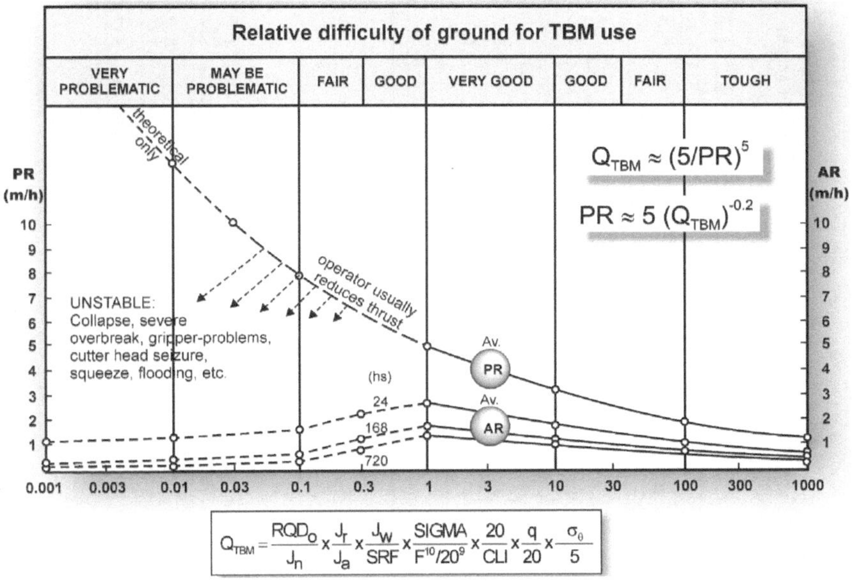

Figure 53. The theoretical (and empirical) basis for the Q_{TBM} calculations, using the program 'Qtbm' (Barton and Abrahao 2003).

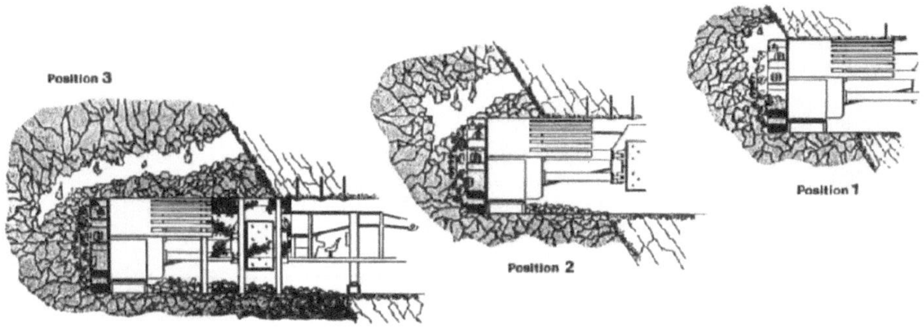

Figure 54. Without probe drilling it is easy to be optimistic, especially when making world record speeds on other sections of the project (Robbins, 1982). Probe drilling is not usually done under the invert where first warnings would be detected, in this and many other cases.

Roald (in Barton et al. 2000/2001) has shown that time and cost of tunnelling are strongly correlated to *Q*-values when the *Q*-value is less than about 1.0, in fact just the same area of sensitivity to *Q* shown by TBM *deceleration gradients (-m)* (see figure on previous page). (The sensitivity to *Q* actually begins at about Q = 10 where support increases begin).

So if the effective *Q*-value can be improved by pre-grouting – in the case of both drill-and-blast and TBM tunnelling, the greatest benefit will be achieved where the Q-versus-cost and Q-versus-time curves are steepest (about $0.01 < Q > 1.0$).

Are long tunnels faster by TBM?

It was shown in Figure 17 that 'central' rock qualities are required for TBM to be significantly faster than drill-and-blast. Figure 17 is repeated here as a reminder.

As tunnel length increases, this 'centrality' of rock quality becomes more important due to the deceleration of advance rate with time, and therefore with tunnel length.

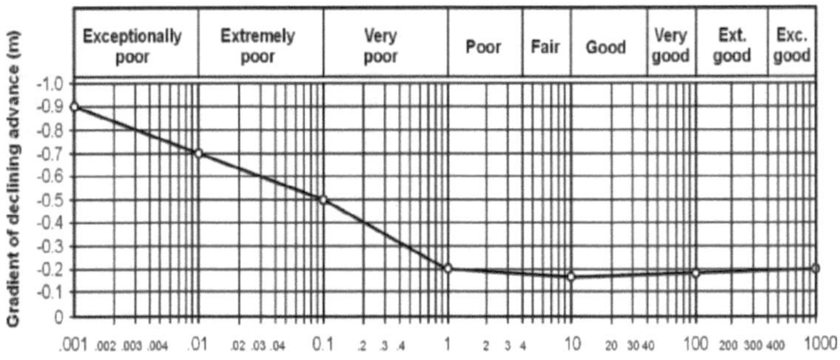

Rock mass quality $Q = \left(\frac{RQD}{J_n}\right) \times \left(\frac{J_r}{J_a}\right) \times \left(\frac{J_w}{SRF}\right)$

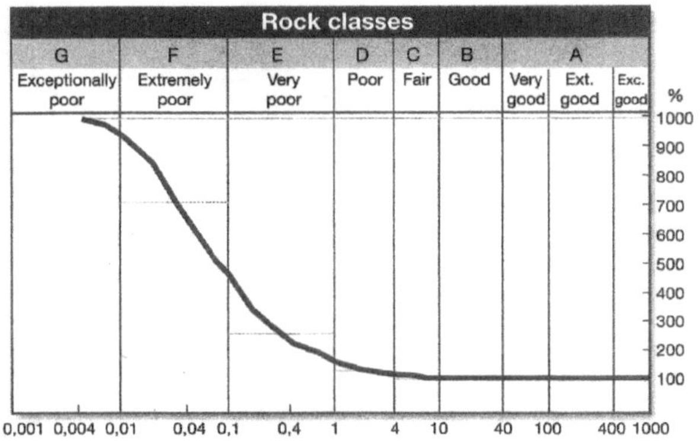

Figure 55. Relative time used in drill-and-blast tunnelling (support included), in relation to *Q*-value (Barton et al. 2000/2001). (Relative cost curve is similar, but the vertical axis extends to 1200%).

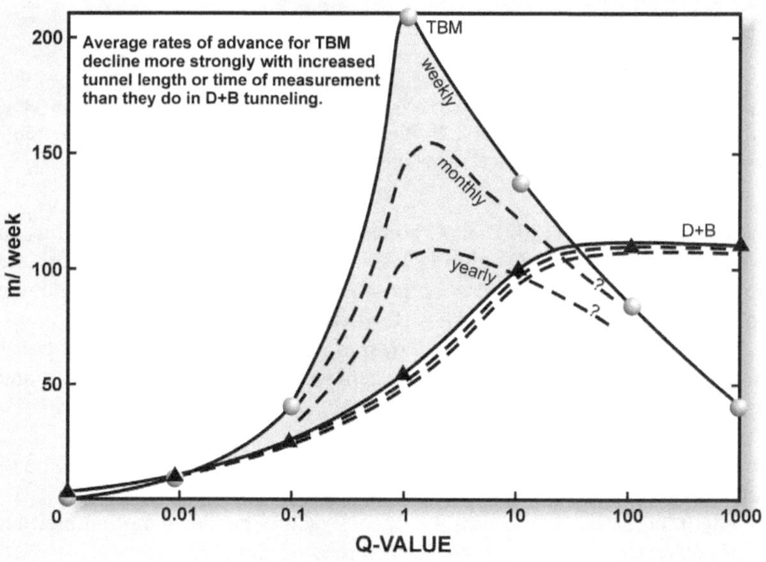

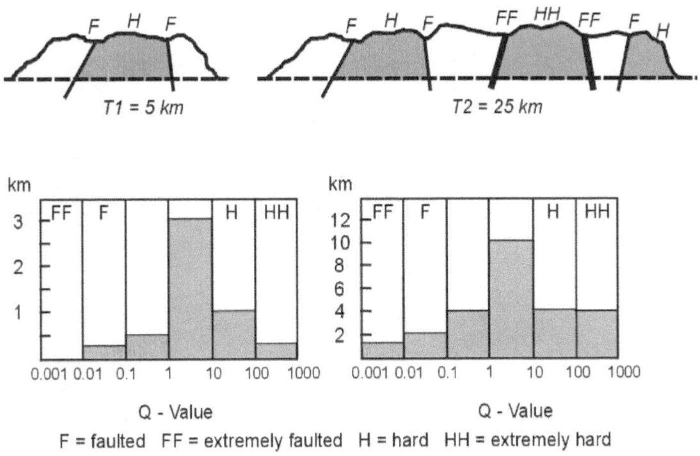

Figure 56. One should not blindly assume that long tunnels are faster by TBM. The longer the tunnel, the more likely that 'extreme value' statistics (of rock quality) will apply, due to a 'large scale' Weibull theory i.e. more 'flaws' the larger the 'sample' (Barton 2002).

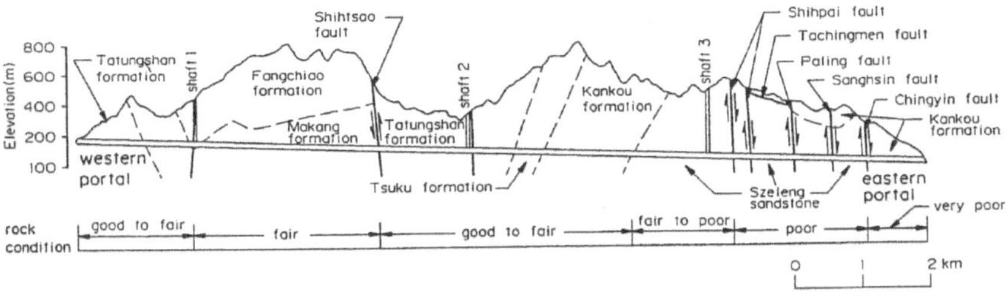

Geologic profile along the Pinglin Tunnel

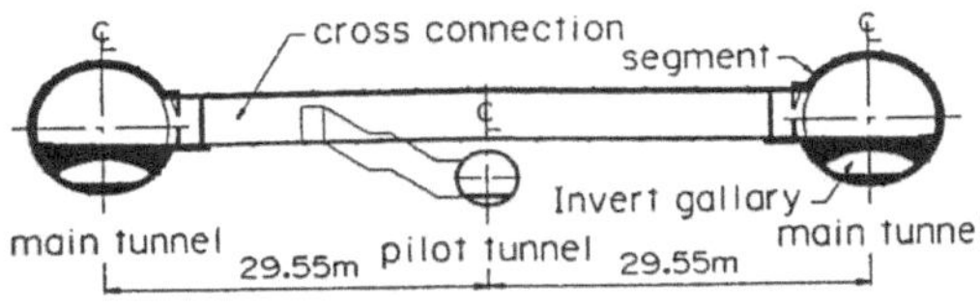

Cross-sectional layout of Pinglin tunnels

Figure 57. Geological section along Pinglin Tunnel, and cross-sectional layout of the three parallel tunnels (Project brochure).

10 EXAMPLES OF A LONG TUNNEL THAT DID NOT GO FASTER BY TBM

Pinglin Tunnel in NE Taiwan is an example of a TBM tunnel (actually three parallel tunnels) where serious faults caused such large cumulative delays, that drill-and-blast 'rescue' from the other end was essential for completion, after some 12 years of struggle to drive this 15 km long twin-road tunnel.

Figure 58. The Eastern portal in 1995, with the first running tunnel TBM under preparation, following advance excavation of part of the pilot TBM, which soon ran into stability problems (Photo: N. Barton).

Figure 59. The impressive Wirth TBM for Pinglin; two machines that eventually met rockmass conditions that frequently defeated them (Wirth Co-advert).

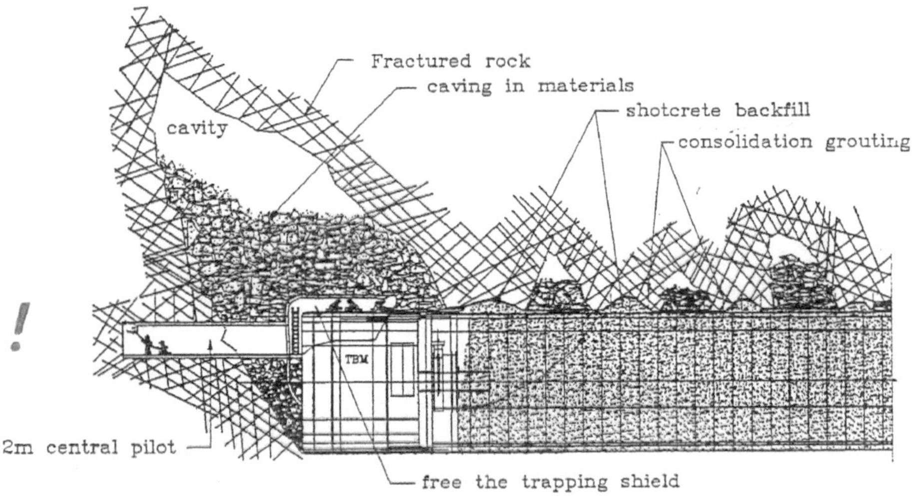

Shows risky means to free TBM cutter head or shield

Figure 60. Graphic illustration of a by-pass situation for one of the Wirth TBM. The TBM was used to cut the bench material for significant lengths of problem ground, following the advance of a drill-and-blast top-heading (Shen et al. 1999).

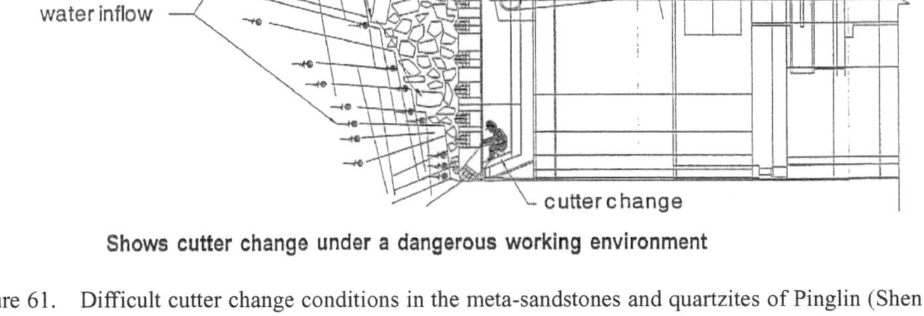

Shows cutter change under a dangerous working environment

Figure 61. Difficult cutter change conditions in the meta-sandstones and quartzites of Pinglin (Shen et al. 1999).

Figure 62. A top-heading in advance of the stuck TBM. Note new breast-plating which was rapidly worn down by the abrasive conditions (Photo: C. Fong).

Figure 63. The 12th by-pass situation for the pilot TBM in 2002 (Photo: C. Fong).

Figure 64. This is the running tunnel where earlier, one of the two 11.7 m diameter TBM had been destroyed in a fault zone. At this 2002 location the actual face of this top-heading is 100 m in front of the present 'face' due to a huge (and fatal) inrush of rock, clay and water (Photo: C. Fong).

11 CONCLUSIONS

1. Fault zones represent the 'Achilles heel' of TBM because, if sufficiently serious, they present the contractor with a situation where the TBM itself is actually 'in the way' of the most efficient pre-treatment or recovery methods that are usually available.
2. Fault zones are a form of 'extreme value' in terms of characterization or classification of the degree of difficulty (and support needs) that they represent. They therefore lie far outside the ideal 'central' qualities where TBM give advance rates that are much superior to those of drill-and-blast tunnelling.
3. Because TBM slowly decelerate as time and tunnel length increase, it is even more important that the rockmass has mostly 'central' qualities. So when a TBM is chosen 'because the tunnel is very long and needs to be driven fast', the opposite may actually occur, as extreme value statistics of rock quality are more likely to be encountered in a long tunnel, which possibly has high over-burden and reduced pre-investigation as a result.
4. Extreme values of rock quality, that may be 'enhanced' by the tunnel length being too long for the choice of TBM, include larger fault zones, higher water pressures, harder or more abrasive rock, and squeezing (or eroding) conditions in fault zones because of high over-burden (or high water pressures).
5. Double-shield machines, with PC-elements for both support and thrust (while re-setting grippers), have been claimed by some as the answer to 'all' variable rock conditions. Such a solution , at a significant extra cost, certainly produces a minimal deceleration gradient, in terms of minimising the slowing advance rate with increasing time or tunnel length, when the rockmass conditions are not extreme. When conditions are extreme, as when stuck in a significant fault zone, the time to recover and pre-treat the ground will tend to be longer, due to the now adverse total lengths of the (double) shields.

Figure 65. Two illustrations of (roughly-speaking) '$Q = 0.001$' and '$Q = 1000$' 'rockmasses', that emphasise the advantage of a classification and characterization method that shows suitable numerical differentiation (1/1 000 000) of extreme qualities, when there is indeed a big difference in their properties (e.g. deformability, shear strength, 'compressive strength'). It is easier to develop simple equations that correlate Q with such properties, than is the case for classification systems that display only 1/10 or 1/20 numerical differentiation in such extremes, as for RMR and GSI (Photo: N. Barton).

6. TBM tend to get stuck when several 'predictable' events combine into an unpredictable 'unexpected events' scenario, usually with extremely low Q-values. It is in avoidance of such situations that TBM can most benefit from probe drilling, both downwards and upwards, and preferably to both sides as well.
7. A degree of preparedness for approaching 'no-longer-upredictable-unexpected-events', can stimulate the use of drainage and systematic pre-injection, which is believed to effectively improve many (or all) of the six Q-parameters, thereby making advance less hazardous.

8. Beware of the effect of compaction or tunnel depth causing an increase in the seismic velocity, if using seismic profiling ahead of a (**TBM**) tunnel. The recording of a reasonable velocity of say 4 km/s may mask actual fault zone qualities, which would reveal a 1.5 to 2 Km/s velocity if encountered nearer the surface.

REFERENCES

Barton, N. 1995. The influence of joint properties in modelling jointed rock masses. Keynote Lecture, 8th ISRM Congress, Tokyo, Rotterdam Balkema: 3: 1023–1032.

Barton, N. 1996. Rock mass characterization and seismic measurements to assist in the design and execution of TBM projects. *Proc. of 1996 Taiwan Rock Engineering Symposium*, Keynote Lecture: 1–16.

Barton, N. 1999. TBM performance estimation in rock using Qtbm. Tunnels and Tunnelling International, Sept. 1999, 30–34.

Barton, N. 2000. *TBM tunnelling in jointed and faulted rock*. 173 Rotterdam: Balkema.

Barton, N. 2001. Are long tunnels faster by TBM? *Proc. Rapid Excavation and Tunnelling Conf. RETC, San Diego, USA*. Hansmire & Gowring (Eds). Soc. Min. Eng.: 819–828.

Barton, N. 2002. Some new Q-value correlations to assist in site characterization and tunnel design. *Int. J. Rock Mech. & Min. Sci.* Vol. 39/2: 185–216.

Barton, N. 2003. TBM or drill and blast. *Tunnels and Tunnelling International*, Jun. 2003: 20–23.

Barton, N. and Abrahao, R. 2003. Employing the Q_{TBM} prognosis model. *Tunnels and Tunnelling International*, Dec. 2003: 20–23.

Barton, N. & de Quadros, E. 2003. Improved understanding of high pressure pre-grouting effects for tunnels in jointed rock. *Proc. of 10th ISRM Congress, South Africa*, Vol. 1, 85–91.

Barton, N. & Warren, C. 1996. Rock mass classification of chalk marl in the UK Channel Tunnels. Channel Tunnel Engineering Geology Symposium, Brighton, September 1995.

Barton, N., Lien, R. & Lunde, J. 1974. Engineering classification of rock masses for the design of tunnel support. *Rock Mechanics*. Vol. 6/4: 189–236. Springer-Verlag.

Barton, N., Buen, B. & Roald, S. 2001. Strengthening the case for grouting. *Tunnels and Tunnelling International*, Dec. 2001: 34–36 and Jan. 2002: 37–39.

Bradley, W.B. 1978. Failure of inclined boreholes. *Transactions of the American Society of Mechanical Engineers*. 101: 232–239.

Deva, Y., Dayal, H.M. & Mehrotra, A. 1994. Artesian blowout in a TBM driven water conductor tunnel in Northwest Himalaya, India. *Proc. 7th IAEG congress, Lisbon*. Oliveira, Rodrigues, Coelho & Cunka (eds) 4347–4354. Rotterdam: Balkema.

Grandori, R., Jaeger, M., Antonini, F. & Vigl, L. 1995. Evinos-Mornos Tunnel – Greece. Construction of a 30 km long hydraulic tunnel in less than three years under the most adverse geological conditions. *Proc. RETC. San Francisco, CA* Williamson & Gowring (eds). Littleton, CO: Soc. for Mining, Metallurgy, and Exploration, Inc. 747–767.

Robbins, R.J. 1982. The application of tunnel boring machines to bad rock conditions. *ISRM symposium, Aachen*. Rotterdam: Balkema. Vol. 2: 827–836.

Shen, C.P., Tsai, H.C., Hsieh, Y.S. & Chu, B. 1999. The methodology through adverse geology ahead of Pinglin large TBM. *Proc. RETC. Orlando, FL*. Hilton & Samuelson (eds). Littleton, CO: Soc. for Mining, Metallurgy, and Exploration, Inc. Ch. 8: 117–137.

Wanner, H. 1980. Stability problems by fullface tunnel boring. *Fjellsprengningsteknikk. Bergmekanikk/ Geoteknikk*. Trondheim: Tapir Press 25.1–25.7.

Geotechnical Risks in Rock Tunnels – Campos e Matos, Ribeiro e Sousa, Kleberger & Lopes Pinto (eds) © 2006 Taylor & Francis Group, London, ISBN 0 415 40005 8

Geomechanical problems in recent Spanish tunnels

J.M. Rodríguez Ortiz
Polytechnic University of Madrid, Spain

ABSTRACT: The extraordinary development of road and railway tunnels in Spain in the last 10 years has raised a number of problems related to the crossing of difficult and mixed ground, high overburdens and the increased use of mechanised excavation. A review is made concerning problems of conventional and mechanised tunnels in hard rock, as karstic ground, roof collapse, wearing of cutting tools, evaluation of support needs, overexcavation, etc. More specifically are dealt with the problems related to soft rocks, especially the theoretical and practical design of tunnels in swelling and squeezing ground as well as some cases where closure of the section and heaving of the invert occurred in soft, non-swelling rock.

1 INTRODUCTION

The last 10 years have shown an extraordinary development of road and railway tunnels in Spain, partly as a consequence of the new high speed railways (AVE) and a great mileage of new highways as well as the improvement of the existing road network.

The design of these tunnels is related to the mountainous morphology of the country, the strict geometrical specifications of modern transportation routes, the environmental constraints and the previous occupancy of the most favourable land.

Safety requirements and the need for greater capacity have led to larger cross-sections than the traditional ones, in the typical range of 90 to 120 m2, or to solutions of twin tunnels.

Leaving apart the metropolitan subway tunnels constructed in varied ground (Madrid, Barcelona, Valencia, etc.) but mainly of a soft and sedimentary nature, most of the road or railway tunnels occur in medium to hard rock, thus allowing the use of conventional excavation and support methods, mainly the NATM and related methods. Only recently new full section methods such as the ADECCO are being considered for tunnelling in favourable ground.

The large cross-sections needed for road tunnels and their moderate lengths (near 90% of the total length of constructed tunnels corresponds to tunnels shorter than 1 km) sets a severe limitation to the use of fully mechanized methods, such as the TBMs.

This is not the case of the most recent long tunnels for high speed railways where a common design of twin circular tunnels, with inner diameter close to 8.50 m, is being used. The excavation is performed by a single or double shielded TBM, although the EPB type has been also considered. The lining is always of precast concrete segments.

Up to now the three tunnels designed in this way have been:

- Guadarrama tunnels (2 × 28.5 km) (in construction)
- Abdalajis tunnels (2 × 7 km) (in construction)
- Pajares tunnels (2 × 24.66 km) (about to start construction)

Several hydraulic tunnels, more than 7 km long, with diameter from 4 to 6 m are also underway.

Geotechnical problems had to be dealt with not only in these long tunnels but also in many others where special ground conditions were anticipated or became apparent during their construction, very frequently due to the unavoidable limitations of the site investigation or to an overoptimistic evaluation of the properties of the ground.

Some of these cases are commented in the following sections.

2 OCCURRENCES IN HARD ROCK

2.1 *Conventional methods*

It must be said that the problems encountered in hard rock are quantitatively of reduced importance when compared with those occurring in soils or soft rock. Leaving aside the cases of defective support or bad workmanship, the following issues are noteworthy:

2.1.1 *Karst*

Empty karstic voids are a limited problem in tunnelling as their detection is not very difficult (probe borings, geophysical methods and even bolt drilling are good sources of information) and they can be easily filled when they occur below the invert or at the wall level. In case of ancient inactive karsts the cavities are usually left empty, designing the tunnel as a tube, partially connected to the surrounding ground.

In the Soller Tunnel, excavated by drill and blast methods in massive limestones, a roof collapse in the 30 m close to the front revealed a huge, cavern extending for more than 20 m upwards. After cleaning the debris the section was reconstructed by means of employing steel arches and steel liner plates, as formwork to cast a thick vault (2–3 m) of reinforced concrete. After evaluating the risk of rockfalls from the walls of the cavity it was considered that the possible impacts could be withstood by the vault.

In El Bocarro Tunnel of the AVE high speed rail Line Lleida-Barcelona a karstic cavern of approximately 1200 m^3 was encountered. The cavity traversed the tunnel alignment over a distance of 10 m, with an average height of 8 m and a transversal width of about 20 m. As the cavity was accessible it was decided to stabilise its surface with 30–40 cm of shotcrete. A heavy duty support and a strong lining were also applied to the tunnel itself.

In very few cases have modern tunnels have passed through active karsts, with piezometric head. In these cases the main efforts have been directed towards the filling of the voids around the tunnel by grouting or pre-treatment before excavation. Internal barriers and bypass adits can also be employed when the cavities are large enough. In some cases realignment of the tunnel has been a practical solution.

2.1.2 *Roof collapse of hard blocks*

Several accidents (Figure 1) have been reported related to the sudden collapse of very large blocks (with widths approximately equal to the tunnel by 15–20 m in length and 5–12 m in height) after

Figure 1. Roof collapse in the Padornelo Tunnel.

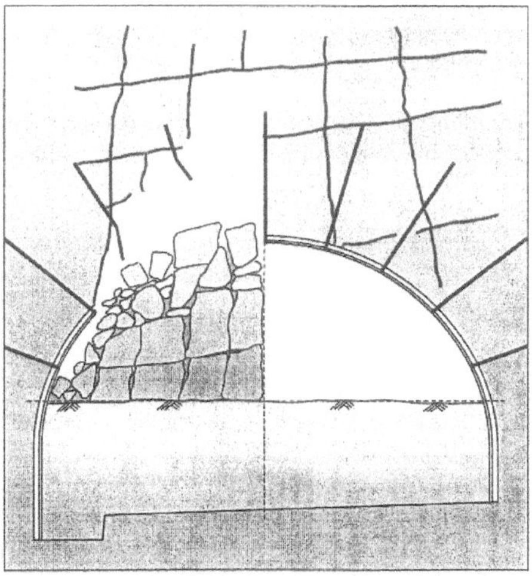

Figure 2. Roof collapse in hard rock.

advancing the excavation face between 3 and 10 m. In all cases the rock was rated as very favourable, with RMR > 75 and a massive appearance. According to the design provisions a very light support was placed, mainly consisting of some bolts and no more than 10 cm of shotcrete.

The cause of the collapse was the presence of near vertical joints, very closed and clean, parallel to the tunnel axis and difficult to detect in the face. The roof was perfectly stable when the joints intersected beneath the tunnel and formed a wedge or key block with no exit possibilities. However geological processes often lead to gradual changes in the joint dip, moving their intersection to a line above the tunnel and defining a wedge free to fall. Obviously the bolts are too short in these cases and the strength of the shotcrete is too low to support the weight of the wedge (Figure 2).

It is not surprising that these collapses occur when the face is advanced several diameters and the rock experiences decompression when, simultaneously, the ground pressures increase.

2.1.3 *Overbreaking*

This is more an economic than technical problem as it involves the need for more support and greater volumes of shotcrete or concrete for filling the out-of-section excavation.

Specifications are becoming more restrictive as refers to the responsibilities of the Contractor on the skilled used of explosives, avoiding overexcavations as well as damaging the rock strength around the tunnel.

In actuality this is not an easy problem to resolve when mixed faces, stratified or brittle rocks, etc. are encountered in the excavation face. Overexcavations depend on a great extent on the rock structure and the type of excavation. Figure 3 shows an empiric relationship between these factors.

In this field the resident engineer or the quality control manager are usually less specialised than the Contractor himself, thus it is difficult to improve the results of defective blasting operations and it is not infrequent to excavate 30–50 cm over the pay-line. Conflicts on this matter are quite common as the Contractor claims that it is impossible to achieve the designed profile, at least with normal methods.

The Agency for High Velocity Railways (GIF) has issued specifications defining the nature of the filling according to the magnitude of the overexcavation (see Figure 4). This originates from the higher price of the shotcrete relative to that of concrete. In the case of tunnel support with steel arches these must be covered by shotcrete, but in case of not using ribs only the nominal or theoretical thickness of shotcrete would be paid, the rest is included as lining concrete.

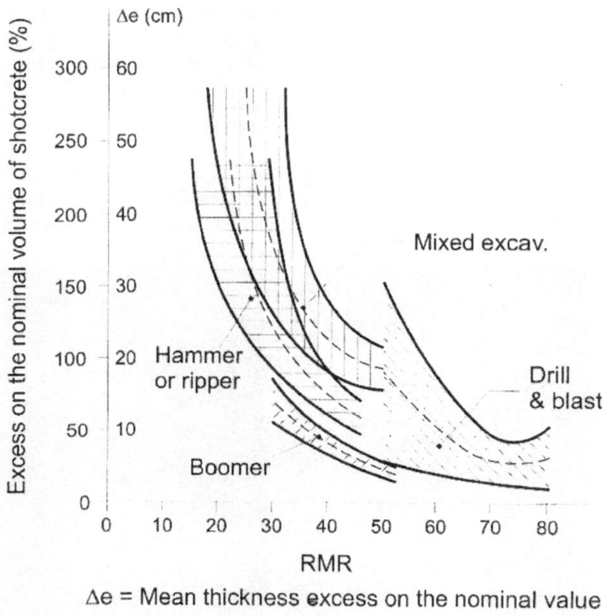

Figure 3. Estimation of overbreak as a function of the rock quality and the excavation method.

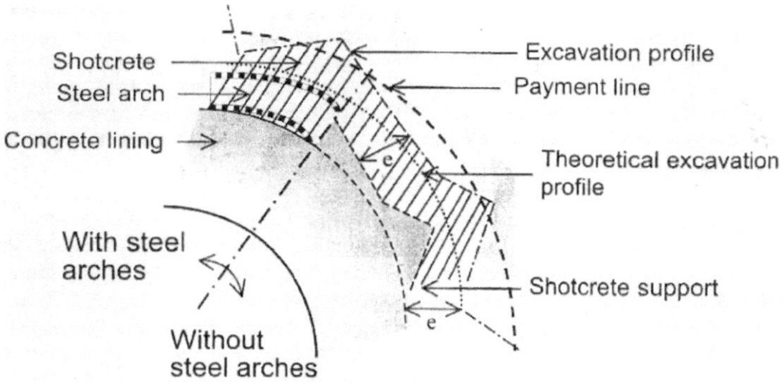

Figure 4. Payment criteria in case of overexcavation.

2.1.4 *Plane roofs*

Flexural bending of stratified plane roofs has also been a cause of some tunnel collapses (Figure 5). The near circular profile of the tunnel is not well suited to this type of ground and, on the other hand, the current designs hardly incorporate the anisotropy derived from the bedding.

2.2 *Tunnels excavated by tunnelling machines*

Although tunnelling machines have been used in Spain since the sixties, they were mainly small diameter TBMs, most of them used in water conveying tunnels. Some shielded machines were also used in the Madrid Metro with limited success. A long lapse of time of approximately 25 years occurred until tunnelling machines come again into operation for long tunnels with large sections above 64 m^2. The biggest effort corresponded again to the Madrid Metro, with 5 EPBs

Figure 5. Fall of a domed block in a stratified roof.

Figure 6. Model of the reaming machine used in the Paracuellos Tunnel.

of 9.40 m in diameter starting to work in 1997. With regard to hard rock machines a new era was initiated in 2000 with the Guadarrama tunnels, in connection with the new high velocity railway lines.

It is interesting to note the case of the Paracuellos tunnel, in the AVE line Madrid-Zaragoza. This tunnel is 4.7 km long and passes through highly tectonized Palaeozoic and Precambrian shales, schists, quartzites, sandstones, and siltstones. The tunnel was started by the drill & blast method but, at the same time it was deemed convenient to drill a pilot tunnel 4.70 m in diameter by means of a small TBM, in order to investigate in advance the ground conditions. This pilot tunnel, 3.4 km in length, would be subsequently widened by a special TBM in order to reach the final diameter of 12.47 m. Due to the poor quality of the ground, 36% of the final length was excavated by conventional drill & blast techniques, with the mechanised solution being limited to 2.95 km (Figure 6).

Mechanized tunnelling has been not yet used for road tunnels due to the large cross-sections required, however, after developing EPB machines 15 m in diameter for the new Madrid inner ring road it is also anticipated that large diameter machines can be used for two or three lane vehicle tunnels.

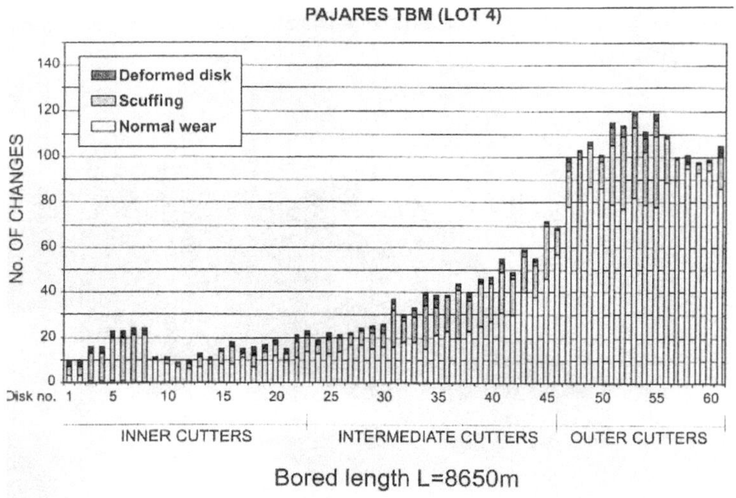

Figure 7. Wear of disk cutters in the Guadarrama Tunnels.

Several types of problems have been reported in recent times:

2.2.1 *Excessive wearing of cutting tools*

This is not a new problem as significant wear is always related to highly abrasive rocks, with high quartz contents. A recent example is the case of the Guadarrama tunnels, in gneissic and granitic rocks, where changing of disk cutters is required every one or two days (Figure 7).

2.2.2 *Collapse of the front*

This problem is mainly associated with open face or shielded TBMs when encountering faults, weathered layers, or soft zones. Rockfalls in the cutterhead area are also frequent in highly fractured rock or when working under high stresses. In soft zones the unstable front is easily excavated which usually results in caving in of the face and the adjacent roof, thus forming voids which may develop into a chimney. The arching of the ground around the void increases the pressures away from the front, mainly on the shield, with a high risk for the machine of being trapped (Figure 8).

One incident of this type occurred in July 2003 in the tunnel of the Line 9 of the Barcelona Metro. The double shielded TBM advanced in schists and hornfels of the Cambrian-Ordovician when a significant overbreak was detected. Although two-thirds of the section consisted of oxidized hornfels the rest was occupied by a weathered layer similar to a sandy clay. The overbreak evolved to important rockfalls and squeezing of the fine material. The cutterhead became trapped and the advance was no longer possible.

Probe drilling showed about 12 m of low strength ground ahead of the face, composed of highly weathered hornfels and porphyry.

When studying the possibilities of intervention it was proved that the rotation of the cutterhead was impeded by the material accumulated at the lower part of the same, instead of the debris fallen on the top of the wheel. This second effect would prevent the use of foams in order to reduce the friction of the shield as the collapsed debris under the liquid injection would have exerted higher radial pressures.

Hand excavation of the material accumulated against the head together with applying the maximum torque of 37.000 kN.m allowed liberation of the machine and to resumption of the excavation. However, in order to prevent further collapses 12.34 tons of cement (14.1 m^3 of grout) was injected, before resuming the advance and 52.9 tons of cement (60.5 m^3) and 192 m^3 of

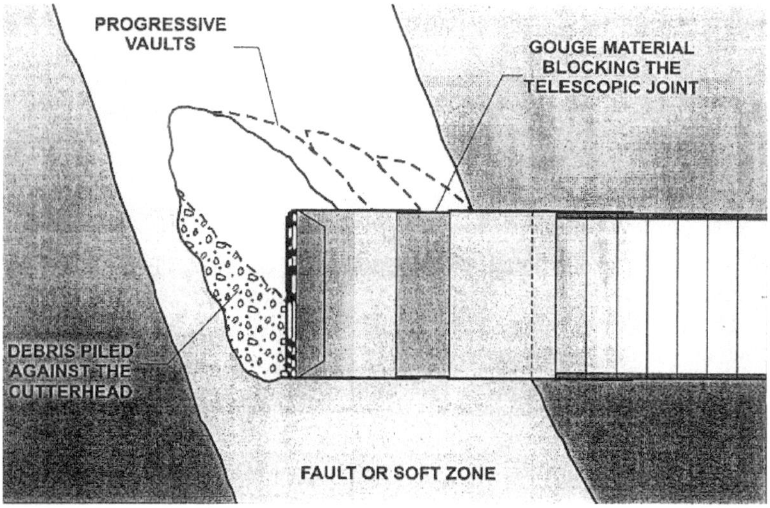

Figure 8. Double shielded machine trapped in a soft zone.

mortar after the passage of the TBM, in both cases from the surface. Injections were also undertaken at the tunnel crown through the TBM itself, amounting 16.4 m^3 of grout.

Another case of ground collapse occurred in February 2004 in the Guadarrama tunnel when overexcavation in an unforeseen weathered granite gave way to an important void and high pressures on the shield, with trapping of the machine. In this case an unfortunate event was the penetration of the ground into the slot left open between the two parts of the double shield.

Although the available unlocking thrust was above 89.500 kN it was not possible to push the machine forward. After a number of attempts, the final solution was to place a steel ring on the last segmental concrete ring and to add more jacks to the existing ones. Thus, after applying a total thrust of about 120.000 kN, the machine could be dislodged and the work resumed.

After a time where the double shielded TBMs were considered as the more appropriate machines for tunnels with high percentage of sound rock, the current point of view is more favourable to very short single shields with high power, in order to overcome the unstable zones.

2.2.3 *Nose diving*

Rock TBMs are very prone to misalignment when working in faces of mixed ground or when they encounter karstic cavities.

The combination of soft and hard ground provokes a deviation of the machine towards the soft side and, when this is in the lower part of the section, the machine can progressively sink until the advance must be stopped in order to realign the machine. When long shields are used the machine works as a cantilever fixed at its rear end, with a marked trend to sink at the cutterhead.

One example has been the deviation of a TBM in the Guadarrama tunnels after penetration into a soft clayey ground. The machine progressively sank along 90 m until reaching a vertical deviation of 1.82 m (Figure 9). Taking advantage of the small cover, a diaphragm of vertical piles, filled with cement mortar, was bored in front of the cutterhead and the ground ahead was reinforced by means of cement injections. After this, a cautious use of the driving jacks allowed a slow realignment of the machine. It took more than 150 m of upward motion to return to the right elevation.

2.2.4 *Karstic cavities*

This problem is more important in mechanized tunnels than in conventional ones, as the machine considerably limits the possibilities of probing ahead of the excavation as well as the restricting

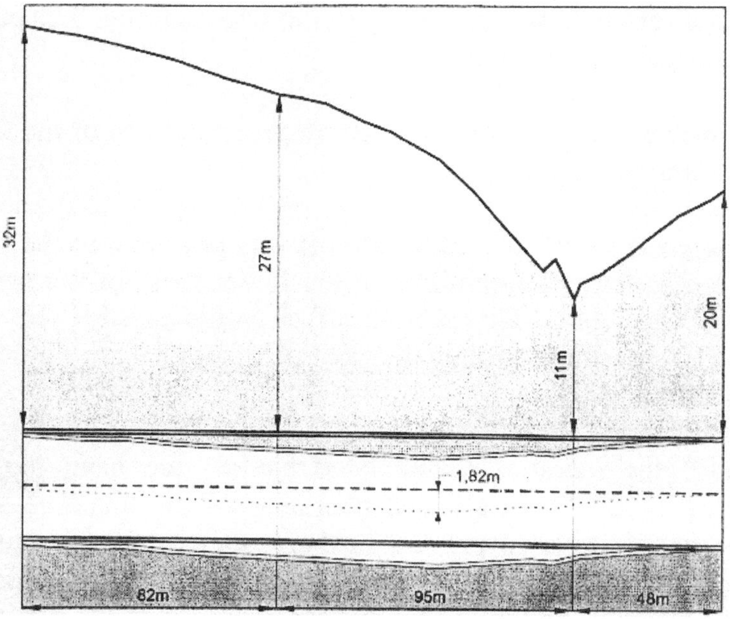

Figure 9. Vertical deviation of the TBM (Guadarrama Tunnel).

space for ground treatment. Drill and blast methods appear to be more flexible in dealing with karstic cavities.

The crossing of karstic cavities with tunnelling machines is impossible without previous filling of the void, at least below the tunnel base. Voids also eliminate the support needed for turning the cutterhead. The detection of the cavity by probe drilling or seismic or sonic logging is very important in order to anticipate the necessary ground treatments. For the time being only moderate karsts have been encountered in mechanized tunnels in Spain and the machines were able to bridge the cavities.

The worst case is when the void is very large (as compared with the length of the head of the machine plus the shield if any) and occurs below the invert of the tunnel. Advance filling is mandatory, in some cases through ancilliary adits.

It is also possible to by-pass the machine in order to construct a conventional tunnel-bridge by means of a canopy of spiles, steel arches and shotcrete, together with a heavy lining of reinforced concrete. Lateral or voids above the crown can be filled with concrete or grouting before or after the passage of the machine.

The presence of voids close to the segmental lining can be assessed by means of radial drilling through the grout holes left in the segments.

Obviously all these problems worsen when crossing active sinkholes or flooded cavities, connected or not to cave systems.

2.2.5 *Water inflows*

Unexpected water inflows may be very harmful to the mechanical and electrical equipment of the tunnelling machines, in addition to the risk to the workers. More often than not these inflows are accompanied by the fines washing out of fines, over break, fall of blocks, etc.

Probe drilling is also required as a preventive measure and, when drainage is not feasible, the only solution is the systematic pre-injection of the water bearing areas.

Obviously these problems can be minimized with the choice of bentonite or mixshields.
In several cases of moderate water inflows the following solutions have been used:

- Pumping of water ahead of the machine from wells drilled from the surface, when the depth of the tunnel is below 50 m.
- Cement grouting from the tunnel
- Injection of polyurethane foams or water-reactive resins from the tunnel.

3 PROBLEMS IN SOFT ROCKS

Obviously a greater number of problems can be expected in soft rocks due to their higher deformability and lower strength but they can be successfully handled with the appropriate supports (shotcrete, bolts, steel arches, etc.) or dividing the tunnel in small sections.

In some cases, however, the design of supports is far from easy due to the difficult evaluation of the forces developed as well as the limited knowledge of the physical phenomena involved. The most striking cases refer to the swelling aspects of some rocks and the squeezing or creep of weak rocks under high overburden.

3.1 *Swelling rocks*

Swelling problems have been reported in tunnels crossing clay- and mudstones, some marly ground and the complex gypsum-anhydrite-clay. Usually these formations are of Miocene or Triassic age.

Although swelling phenomena have been studied in Spain from more than 50 years, the cases related to tunnelling are quite scarce and mainly in the field of hydraulic tunnels.

When the risk of expansive ground was detected in some recent tunnels it became apparent that there was a lack of technical theory available with regard the evaluation of the relevant soil properties as well as reliable computation methods.

The commercially available codes did not include models for expansive behaviour and only simplified methods, mainly based in thermal analogies, were given.

On the other hand, laboratory testing of some of these materials takes a time greatly exceeding the term available for the design or even the construction and, furthermore, the translation to field conditions is far from clear.

The experience gained in Germany in a number of tunnels with long term monitoring as well as the theoretical developments performed by Kovary, Wittke, Steiner, etc. has been of great help.

Even in Germany the proposed solutions are not unique (Figure 10) and some authors postulate the use of very thick inverts in order to counteract the maximum swelling pressures expected whereas others uphold the provision of an "expansion chamber" to allow the swelling of the ground without transmitting significant pressures to the invert. This second solution has been used in the Taubenbloch T8 Tunnel (Steiner 1993) and the Engelberg Tunnel (Kirschke 1998) but it seems of limited practical value due to its very complex construction (Figure 11). On the other hand, the buffer chamber or a compressible layer do not prevent of high swelling pressures in the long term.

The first approach has been followed in Spain in some of the tunnels with this problem. Some variations include the combination of a thick invert with vertical permanent anchors.

From a theoretical point of view the following model, according to an original proposal by Wittke (1979 and 2000), has been used:

Adoption of a stress-swelling law of triaxial type, instead of the customary mean stress or first stress invariant

Assumption of a rheological law time-change of volume

To this respect it was maintained the well-known hypothesis of Huder-Amberg (1970):

$$\Delta V = K - m \log p \tag{1}$$

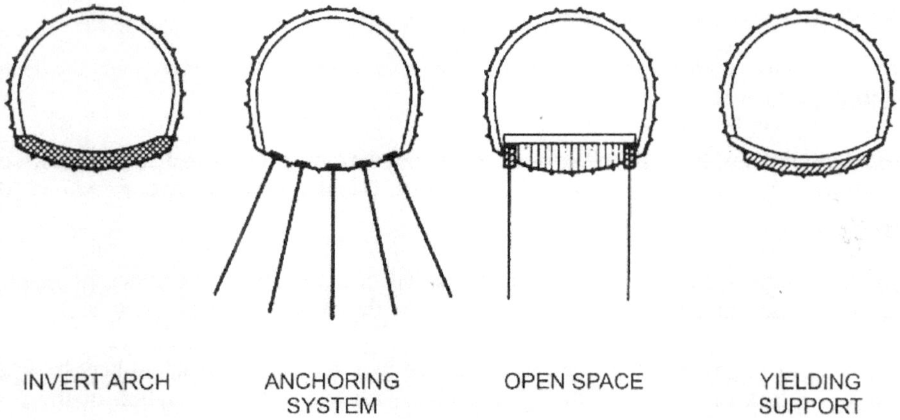

Figure 10. Solutions for tunnel inverts in swelling ground (after Kovari et al. 1988).

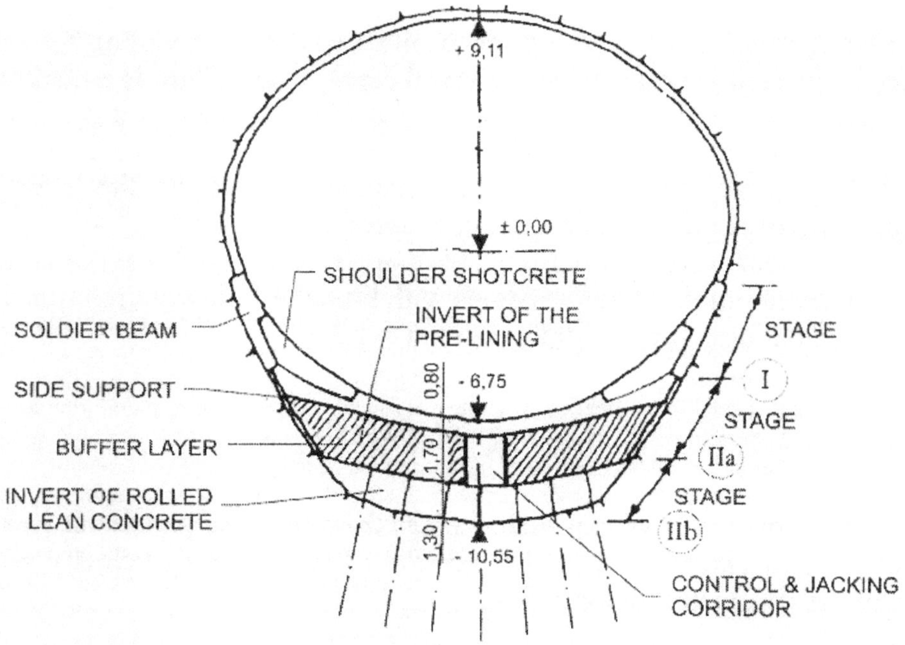

Figure 11. Section of the Engleberg base tunnel in a ground with the anhydrite.

which can also be written as

$$\varepsilon_{z\infty}^{q} = K_q \log\left(\frac{\sigma_z}{\sigma_0}\right) \tag{2}$$

The lower limit of this logarithmic law is the free swell and the upper limit the swelling pressure (Figure 12).

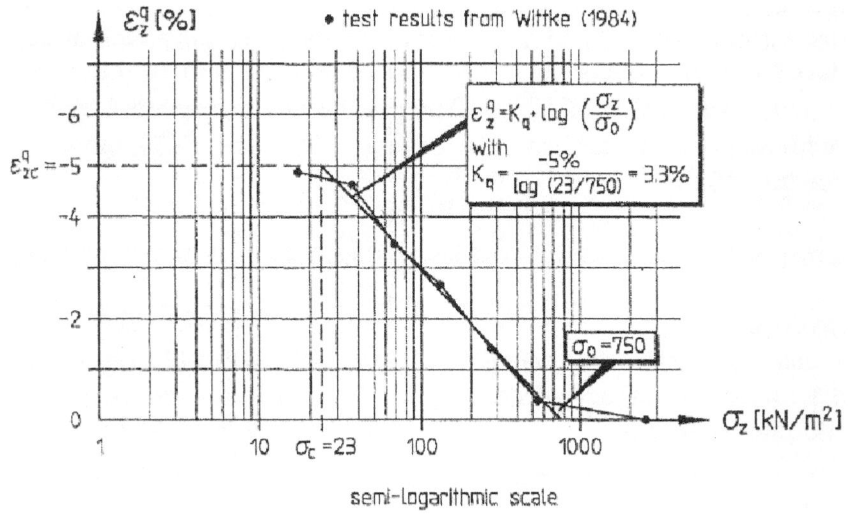

Figure 12. One dimensional swelling law (from Grob 1972 in Wittke 2000).

For the time being it is not possible to introduce anisotropy in the swelling behaviour although this property may be present in some of the observed cases, where there is a marked layering parallel to the tunnel axis.

Coaxiality between the directions of the principal values of the stress tensor and the corresponding ones of the deformation tensor is assumed.

As refers to the development of deformations with time the following relationship is assumed for the three principal directions of deformation $i = 1, 2, 3$.

$$\frac{\partial \varepsilon_i^q(\mathrm{t})}{\partial t} = \frac{1}{\eta_q}[\varepsilon_{i\infty}^q - \varepsilon_i^q(\mathrm{t})] \tag{3}$$

The main difficulty in expression (3) arises from the time parameter η_q which includes the expansion of the material as well as the increase of the swelling velocity with the elastic and plastic parts of the volumetric deformation.

$$\frac{1}{\eta_q} = a_0 + a_{el}\varepsilon_v^{el} + a_{vp}\min\{\varepsilon_v^{pl}, \max EVP\} \tag{4}$$

EVP being the upper limit of the plastic deformation.

The process ends with the complete saturation of the material when enough water is available for fully developing expansivity.

When plastic deformations occur the permeability of the ground enormously increases, along with a degradation of the strength parameters and a significant increase of the deformability. Some materials, however, as those containing anhydrite may show a self-sealing behaviour with regard to their permeability due to the closure of the joints when the matrix expands (M. Wittke 2003).

The computations are quite complex as an iterative procedure must be followed.

This method has been used in the Fabares road tunnels, part of the Autovia del Cantabrico, where the occurrence of anhydrite, gypsum, and swelling clay was anticipated. These Triassic rocks appeared partially covered by breccias, dolomites and limestones from the Lias period, with a maximum overburden of 170 m. The ground was intercepted by a number of minor faults and joints with moderate water head. According to Steiner (1993), the anhydrite only appeared at depths above 90–110 m.

The long-term swelling tests carried out showed important heave strains such as:

- Anhydrite 8% at 1443 days
- Red clay with anhydrite seams >15% at 1413 days
- Black mudstone and anhydrite >11% at 1562 days
- Remoulded anhydrite >30% in 1420 days
- Remoulded red clay with anhydrite seams >65% in 1420 days

These tests confirm the slow saturation process of the anhydrite as well as the harmful effect of the clay interbedding.

The swelling pressure tests were not very conclusive as the samples disintegrate at pressures near 2 MPa.

Once proved the risk of swelling phenomena and with the assistance of Prof. Wittke, it was decided to install a resistant invert designed to withstand pressures up to 4 MPa, although there is some evidence that 6 MPa may be a practical limit of the maximum expected pressures.

Several options were considered for the lining (Figure 13) and these sections were analyzed by means of the model above.

Figure 14 shows the final design of the lining and Figure 15 is a diagram of the stresses in the lining for 4 and 6 MPa swelling pressures. In Figure 16 appears a view of the placement of the reinforcing.

These tunnels were finished in the summer of 2003 and their performance is quite satisfactory.

It has been observed, however, that it is advisable to omit the conventional drainage between the temporary support and the final lining in order to avoid the progressive leaching of the gypsum around the tunnel and the development of cavities or solution channels.

At approximately the same time as the construction of the aforementioned tunnels, there appeared severe distresses caused by swelling in three tunnels of the High Velocity Line Madrid-Barcelona, section Lleida-Martorell.

In two of the tunnels the ground consisted of claystones, marls and red sandstones of Eocene-Oligocene age, whereas in the third tunnel shales with gypsum seams of the Muschelkalk period predominated.

The tunnels were excavated without problems by the drill and blast method. The design anticipated the use of plane or curved inverts according to the nature of the ground as detected along the excavation (Figure 17). This is in fact a difficult task and, in most of the length, it was decided to cast a plane slab.

Soon after placement important heave of the base slab was observed, with the intensity varying along the length of the tunnels but amounting more than 40 cm in some places (Figure 18).

In order to check the characteristics of the phenomenon a curved invert of circular profile was constructed, over a length of 300 m, capable of withstanding a pressure of 0.5 MPa, similar to the design pressure of the rest of the tunnel support. This invert was instrumented by total pressure cells, sliding micrometers, vibrating wire piezometers and heave pegs.

At the same time borings 10 m deep were drilled every 100 m of tunnel and mineralogical and geotechnical tests were performed on the obtained samples. The test showed important contents of anhydrite, dolomite, paligorskite and gypsum. Hence, the heaving effects observed were attributed to the hydration of some clay minerals and to the transformation of anhydrite into gypsum.

The water conditions around the tunnels were quite erratic, varying from dry ground to water levels between 2 m above up to 12 m below the bottom of excavation.

The measurements of the 45 load cells showed a wide range of values. Most of the cells did not record any pressure but 10 cells marked pressures above 1 MPa, with a maximum of 2.5 MPa. The maximum values were always measured at the edges of the invert, not at the centerline.

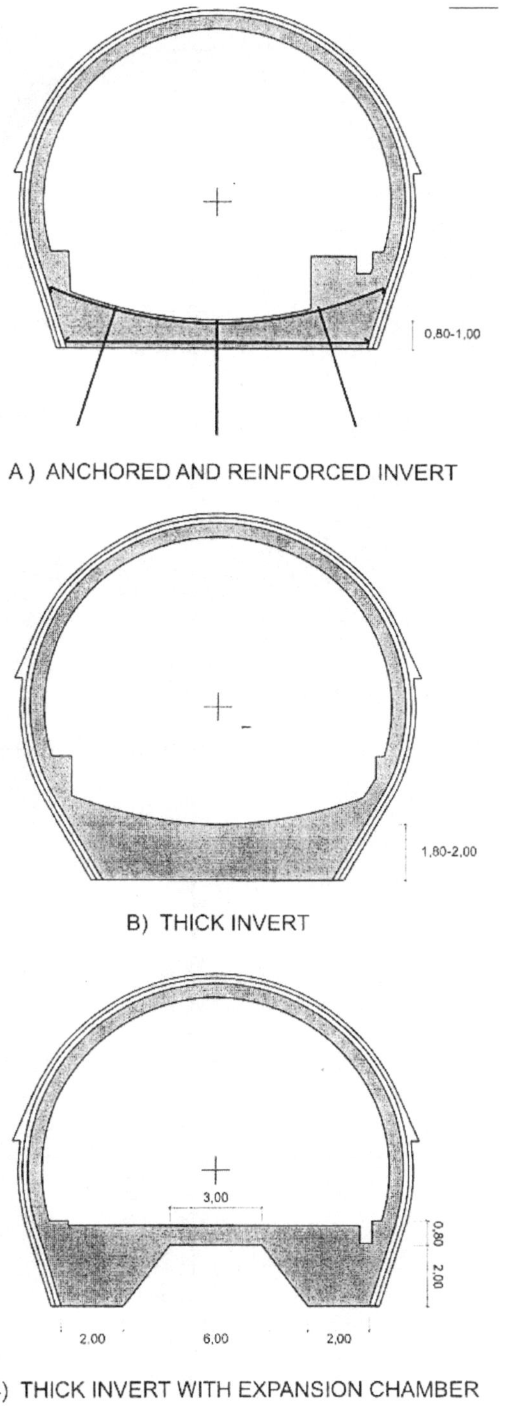

Figure 13. Tested solutions for the swelling sections of the Fabares Tunnel.

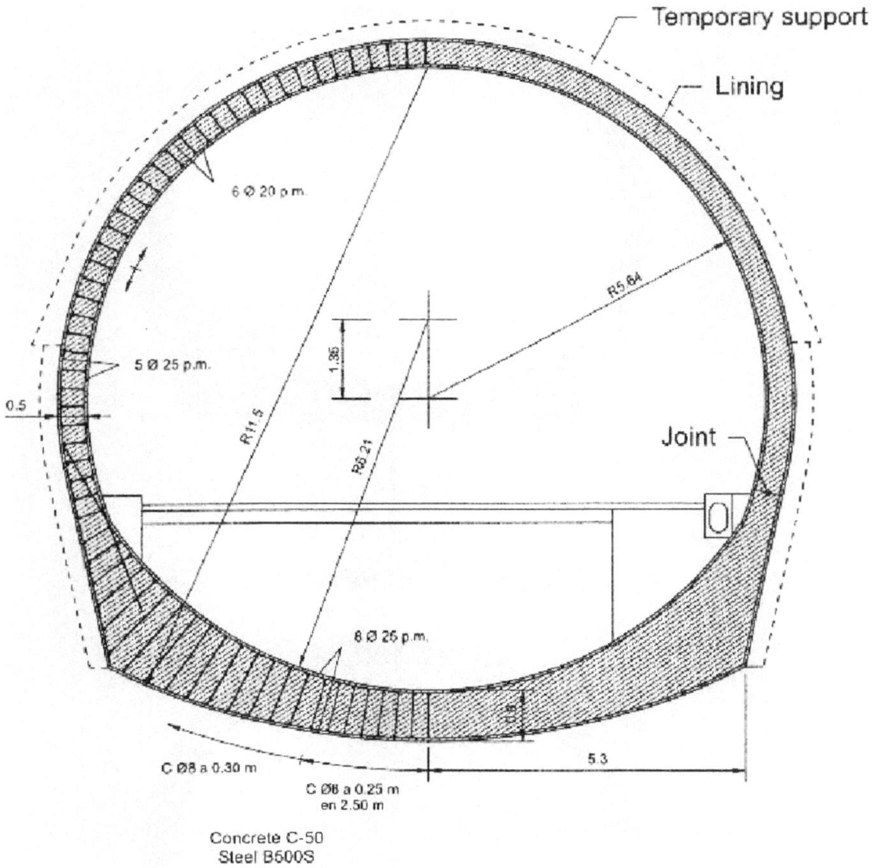

Figure 14. Final section of the Fabares Tunnel.

As the mean geostatic stress was also of 2.5 MPa and this value was also recorded in a similar marl at the Asco Nuclear Plant it was decided to adopt this value as the design pressure of the lining.

The measured "active zone" or depth with heaving strains was of about 4 m, i.e. approximately the 30% of the tunnel width. As expected, the plane slab did not prevent the heave which was linearly increasing upwards. However, close to the circular invert, the sign of the strains was changed to compression, although the invert deflected upwards between 3 and 6 cm.

As a result of pressiometric tests a deformation modulus of 2 GPa has been adopted for the rock mass.

The conclusion of all these studies has been to recommend the placement of a thick circular lining designed for a swelling pressure of 3 MPa at the base.

Figure 20 shows a detail of this lining and its reinforcement. In several sections the existence of the previous temporary support limited the thickness of the lining. Thus several solutions of varying thickness and reinforcement have been used.

Expansive clays, without the presence of anhydrite, have been found in a number of recent tunnels but swelling pressures rarely have exceeded 0.5 MPa (local values of 0.8 to 1.2 MPa are exceptional). Curved inverts with thickness between 40 and 80 cm are the usual solutions, with some enlargement of the lower part of the walls where a high concentration of stresses arises.

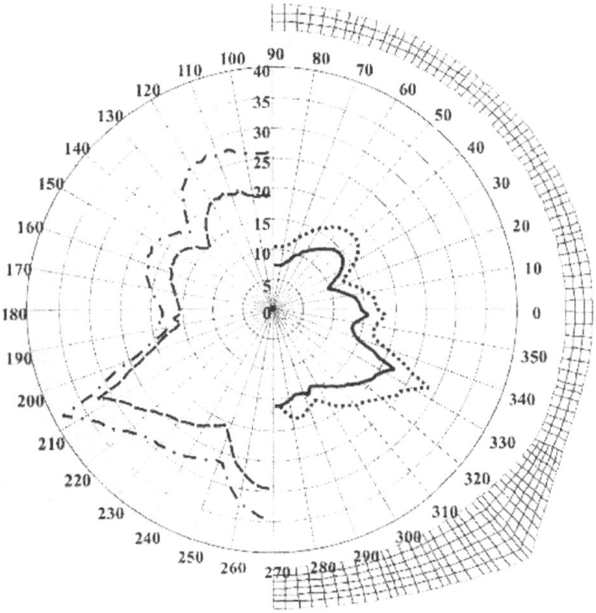

50 years P=4 MPa ...... 200years P=4 MPa —— · 50 years P=6 MPa — · · 200 years P=6 MPa

Figure 15. Maximum compressive stresses in the lining for several pressures.

Figure 16. Reinforcement of the Fabares Tunnel.

3.2 *Squeezing rocks*

This problem is quite new in Spain given that only recently tunnels under high overburden, near 1000 m, have come into consideration. Of course this a well-known phenomenon in the Alpine tunnels and a number or criteria or theories are available (Singh et al. 1992; Aydan et al. 1993; Hoek and Marinos 2000, Barla 2001, etc.). However, as in the case of swelling ground, no theory can be accepted as fully satisfactory .The main problem is the scarcity of long term measurements

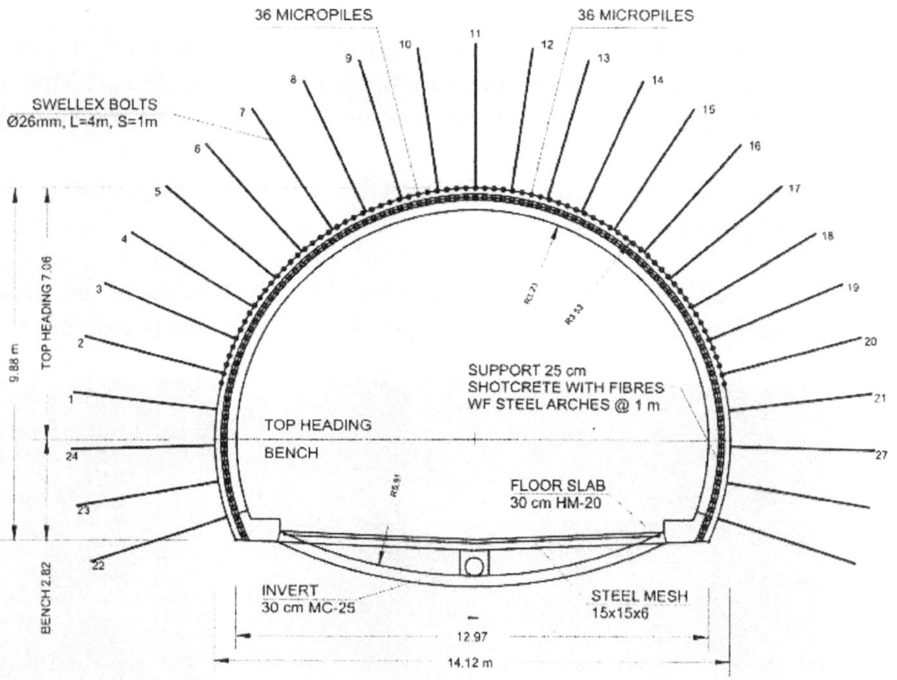

Figure 17. Original design for heavy support requirements (Lilla Tunnel).

which would allow the confirmation of the different predictions. This lack of data is more exacerbated in the case of mechanized tunnels with segmental lining.

Significant advances have been made in rock salt and some specific materials, but these findings are not directly applicable to common weak or hard rocks.

Several models have been used in Spain in order to formulate squeezing phenomena. The most promising one involves an instantaneous elastoplastic behavior and a power-law (Norton 1929), for the creep. The power-law has the following form:

$$\dot{\varepsilon}_{cr} = A\overline{\sigma}^n \tag{5}$$

being $\dot{\varepsilon}_{cr}$ the velocity of secondary creep; A, n coefficients related to the material properties and $\overline{\sigma}$ the Von Mises stress invariant, expressed as:

$$\overline{\sigma} = \sqrt{3J_2} \tag{6}$$

where J2 is the second invariant of the deviatoric part of the stress tensor. The mean stress is defined as

$$s = \frac{1}{3}(\sigma_x + \sigma_y + \sigma_z) = \frac{1}{3}(\sigma_1 + \sigma_2 + \sigma_3) = \frac{I_1}{3} \tag{7}$$

The deviatoric stress tensor s_{ij} is obtained by substracting s from the principal stresses in the stress tensor, i.e.:

$$s_{ij} = \begin{bmatrix} s_x & s_{xy} & s_{xz} \\ s_{xy} & s_y & s_{yz} \\ s_{xz} & s_{yz} & s_z \end{bmatrix} = \begin{bmatrix} \sigma_x - s & \tau_{xy} & \tau_{xz} \\ \tau_{xy} & \sigma_y - s & \tau_{yz} \\ \tau_{xz} & \tau_{yz} & \sigma_z - s \end{bmatrix} \tag{8}$$

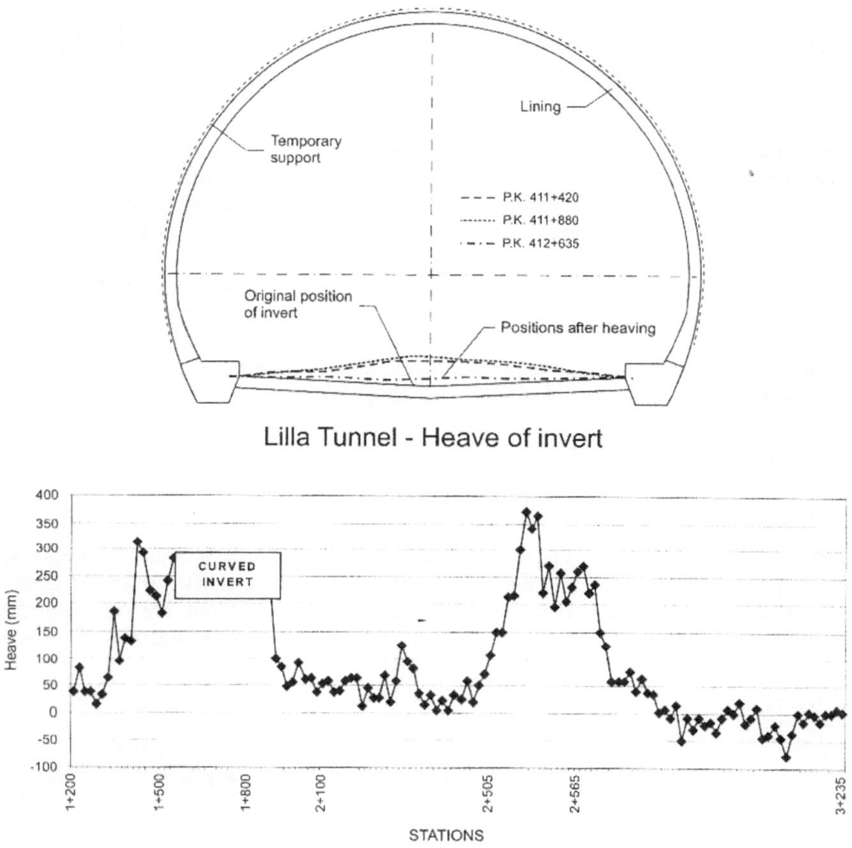

Figure 18. Invert heavings along the Lilla Tunnel.

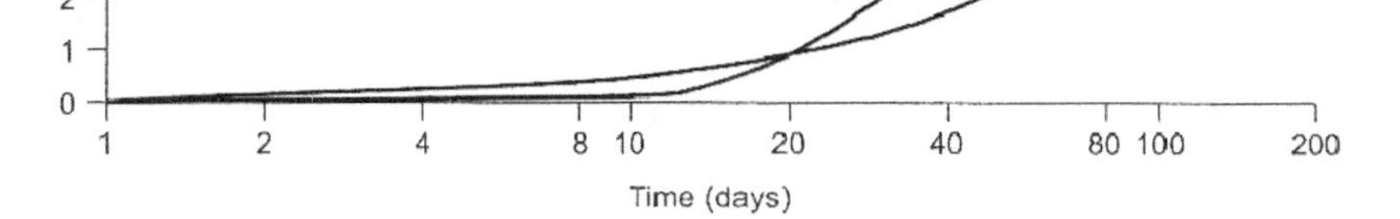

Figure 19. Heave with time of the test section.

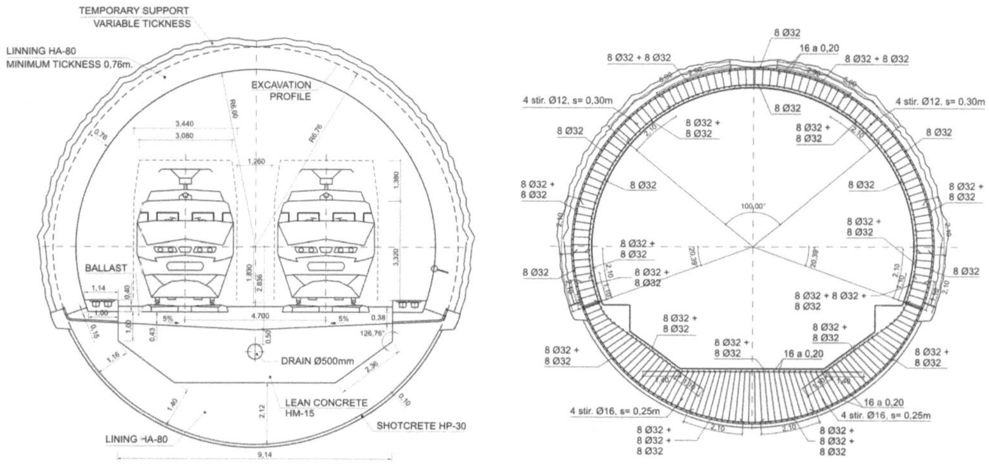

Figure 20. Detail of the reinforced lining (Lilla Tunnel).

Thus

$$J_2 = \frac{1}{2}(s_x^2 + s_y^2 + s_z^2) + s_{xy}^2 + s_{xz}^2 + s_{yz}^2$$
$$= \frac{1}{6}[(\sigma_1 - \sigma_2)^2 + (\sigma_1 - \sigma_3)^2 + (\sigma_2 - \sigma_3)^2] \tag{9}$$

Finally the Von Mises invariant is obtained:

$$\overline{\sigma} = \sqrt{\frac{3}{2}(s_x^2 + s_y^2 + s_z^2) + 3(s_{xy}^2 + s_{xz}^2 + s_{yz}^2)} \tag{10}$$

The viscoplasticity is modeled by combining this viscoelastic model with one of the usual plastic models such as the Mohr-Coulomb or the Drucker-Prager's. This last model is defined by:

$$\tau = k_\phi + q_\phi s \tag{11}$$

where $\tau = \sqrt{J_2}$ and s = mean normal stress as defined above; k_ϕ; q_ϕ are material properties which can be related to the cohesive and frictional parameters of the Mohr – Coulomb model:

$$k_\phi = \frac{6}{\sqrt{3}(3 \pm \sin\phi)} c \cos\phi \tag{12}$$

$$q_\phi = \frac{6}{\sqrt{3}(3 \pm \sin\phi)} \sin\phi \tag{13}$$

With the assumption:

$$\sigma_2 = \frac{\sigma_1 + \sigma_3}{2} \tag{14}$$

strictly valid for circular cavities in elastic media, together with the Mohr Coulomb criterion expressed as

$$\sigma_1 = \sigma_c + N_\phi \sigma_3 \tag{15}$$

the formulae above are simplified to

$$q_\phi = \sin\phi \tag{16}$$

$$k_\phi = c\cdot\cos\phi \tag{17}$$

It is generally accepted that the creep is activated when the stresses reach values above a certain percentage of the peak strength of the material, being frequent values around 50–70%. (Figure 21).

Following dilatometer tests in Mol clays (Belgium), Rousset (1988) has observed said creep onset and he has proposed the viscoplastic model shown in Figure 22.

This model takes the end deformation as the sum of an instantaneous component and a long-term de-formation due to creep:

$$\varepsilon = \varepsilon_i + \varepsilon_{cr} = (\varepsilon^e + \varepsilon^p) + \varepsilon_{cr} \tag{18}$$

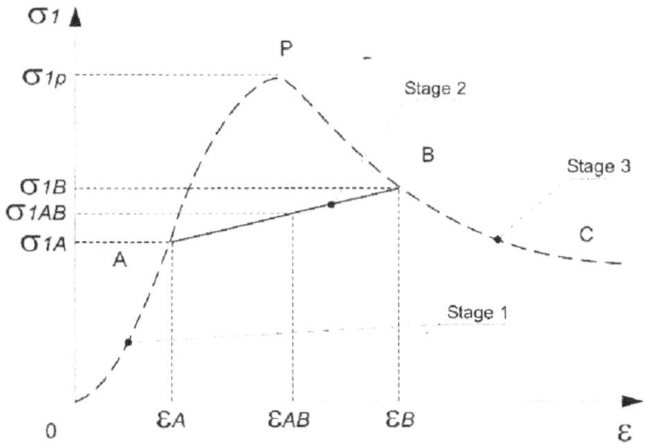

Figure 21. Onset and development of creep, as adopted in the Pinglin Tunnel (From Barla 2001).

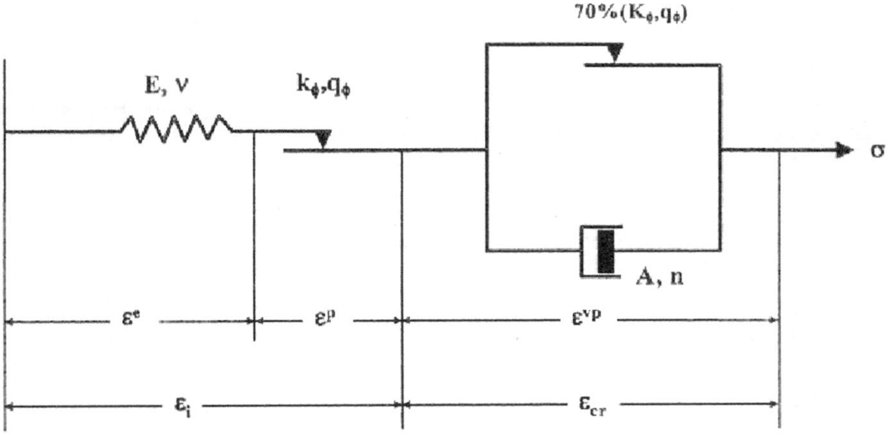

Figure 22. Viscoplastic model according to Rousset (1988).

The instantaneous deformation is represented by the spring and the friction glider whereas the creep deformation is represented by a dashpot in parallel arrangement with a rigid-fragile glider which limits the action of the dashpot to levels of stress above a certain fraction a of the peak strength, for instance

$$\dot{\varepsilon}_{cr} = \begin{cases} A\bar{\sigma}^n & >70\% \text{ strength} \\ 0 & \leq 70\% \text{ strength} \end{cases} \tag{19}$$

Other models rely on the estimation of the long-term stresses, irrespective of their development with time. The increase of the loading on the lining is obtained through degrading the mechanical properties of the ground in the plastic zone around the tunnel. Estefania (2002) has proposed the following values for the degraded parameters

$$C_{cr} = c'/3$$
$$\phi_{cr} = \phi'$$
$$E_{cr} = 1/2E_0$$

Other authors postulate a reduction to zero in the effective cohesion although the friction is maintained at its original value. This may be correct in clay soils or similar but in weak rock a structural degradation can be expected, leading to a reduction on the original friction angle. The problem is to quantify the de-formation required to achieve this degraded condi-tions (strain softening) although a 1% value is often quoted as a plausible value. This establishes a clear distinction between tunnels made by conventional methods, with large convergences before application of the lining and those using tunnelling machines, where the segmental lining is placed with limited closure of the section.

These non time-dependent models are useful for evaluating the final stresses on the linings but give no indication as to the stress-strain behaviour during other stages of the work, for instance the pressures on the shields during eventual stoppages. These pressures can result in a trapping of the machine if not enough power is available to free it.

Figuer 23 shows the expected pressures on the lining versus time, as computed for the Pajares tunnel. These results correspond to a depth of 900 m in the Oville formation (Carboniferous shales), with the following properties: $E = 2400$ MPa; $\nu = 0.3$; $c = 0.60$ MPa; $\phi = 22°$; A = 5E-28, and $n = 2.7$

The length of the shielded TBM is of 9 m and the segments are 0.50 m thick. High quality concrete with strength above 80 MPa is required.

Figure 24 shows the thrust to be applied by the machine for advancing under the expected pressures. A friction coefficient of 0.25 has been adopted. During continuous advance a moderate thrust is necessary, but in the case of a stoppage the longer the duration of the stop the higher will be the required thrust to resume operation. A week seems a reasonable period for an accidental stoppage but this is a very difficult prediction.

The problem of squeezing ground can be tackled in two ways as follows:

- Using conventional methods, allowing as much deformation of the ground as possible before placing heavy supports or the final lining (Figure 25). Due to the difficulty of bolt installation and the risk of cracking the shotcrete if placed too early, only the use of sliding steel ribs seems of practical use as primary support. Special methods such as leaving gaps in the shotcrete lining are difficult to implement and control.
- Excavation by means of shielded machines with immediate placement of concrete segments. The overcutting must be as high as possible, leaving a gap outside the segments in the range of ≥30 cm for machines over 9 m in diameter. Even these huge gaps are unable to absorb the necessary convergences, thus the use of thick segments with high strength concrete or even cast iron is required.

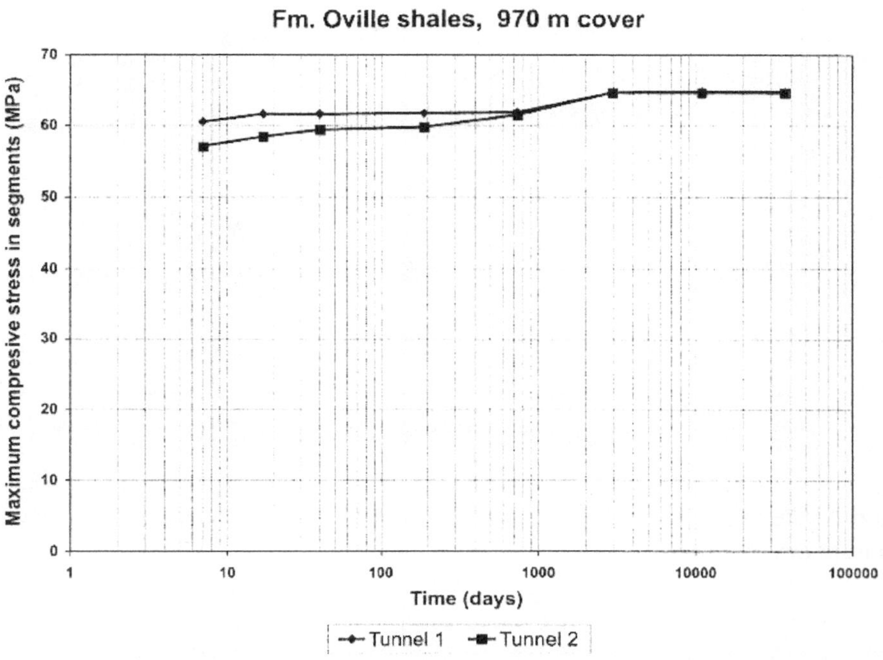

Figure 23. Stresses in the segments versus time.

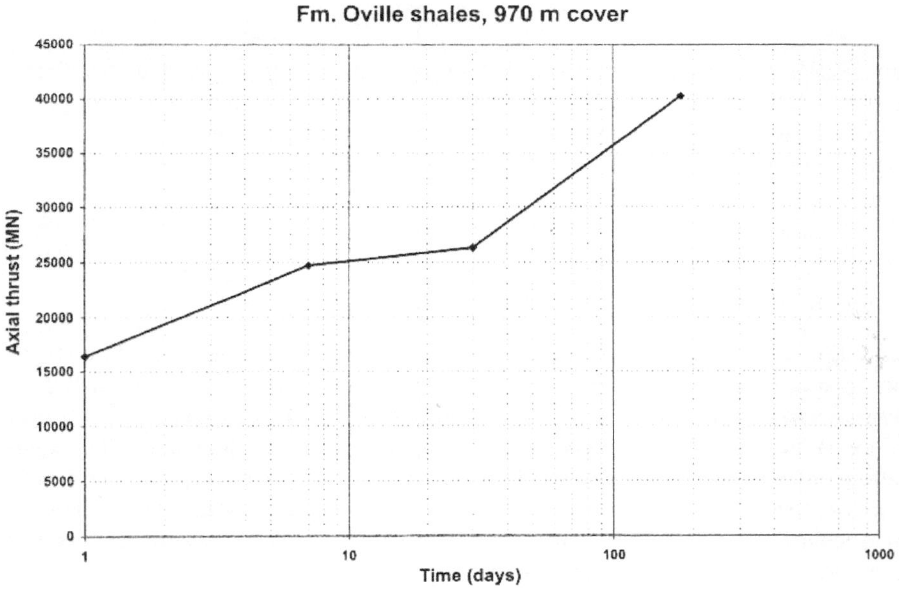

Figure 24. Thrust required for advancing the shield versus stop-page time.

Obviously the second method is preferred but is difficult to justify when the risk of squeezing is concentrated over a short length of tunnel. On the other hand the conventional methods are still preferred to TBM operation due to their higher flexibility and when the risk of the machine becoming trapped is high.

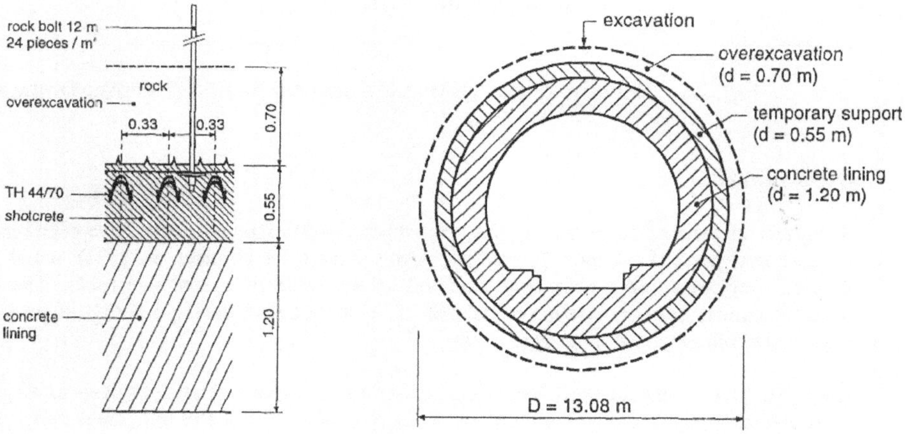

Figure 25. Temporary and permanent linings and cross-section for large convergences in the Gothard Base Tunnel (after Kovári and Ehrbar 2000).

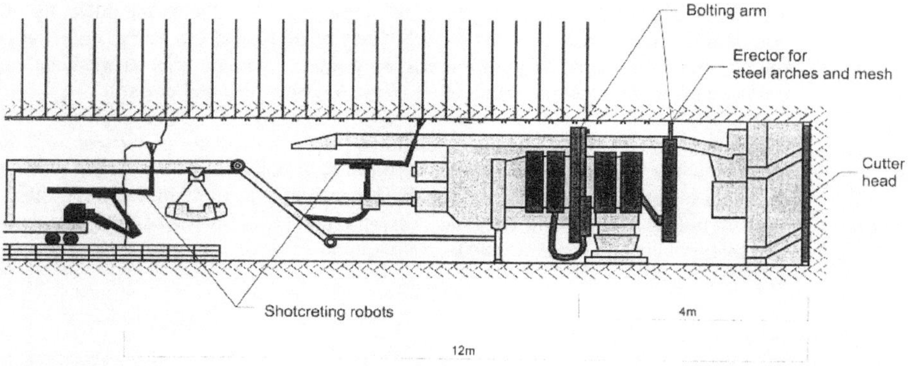

Figure 26. Scheme of the TBM in the original design (Pajares Tunnels).

The use of compressible elements placed in the joints of the segmental lining or shields with parallel blades supported by hydraulic rams being able of accommodating some radial deformations are not being taken into account, at least for the time being.

Following some sections of the Gothard Base Tunnel, the original design for the Pajares Tunnel involved an open TBM with facilities for installing bolts and shotcreting close to the face (Figure 26). However all the bidders preferred a short shielded TBM with segmental lining due to the difficulties expected with the other method. However, due to the uncertainties associated with the predictive models the final construction method remains open. In the case that the field measurements confirm the predictions, the machines would be stopped and the tunnel continued by conventional methods through the problematic sections, leaving an enlarged profile in order to allow the towing of the machine until the new section of good ground is reached.

3.3 *Other problems*

A number of cases of invert heave, not related with swelling phenomena, have been reported in the last few years (La Canda, El Padrún, Ateca, Fuente La Higuera etc.). The usual features involve near-planar floor slabs and weak mudstone, shale or marls type rocks. The heaving occurred shortly after completion of the slab or, in some cases, during a period of several years after commissioning and opening to traffic.

Although several explanations have been presented to explain these occurrences (Figure 27) the most plausible appears to be the buckling or hogging of the, normally thin, slabs under the great compressive forces exerted by the side walls of the tunnel due to delayed effects such as

- The increase of water pressures around the tunnel attributable to defective drainage.
- The progressive softening of the rock near the foot of the side walls and below the slab.
- Long term heave at the tunnel base due to unloading caused by the excavation itself.
- Creep failure of the overstressed lower bolts, too short to anchor outside the plastic zone.

Other effects such as the uplift pressures of the recovering groundwater conditions or the softening of the ground below the slab due to the cyclic traffic loads have been also reported.

Obviously these problems could have been avoided through a stronger invert, of curved shape, and/or longer anchors near the foot of the tunnel walls.

In the reported cases the solutions involved:

- Insertion of longer and heavier anchors in the walls. Vertical anchoring of the floor slab.
- Improvement of the drainage in order to reduce water pressures.

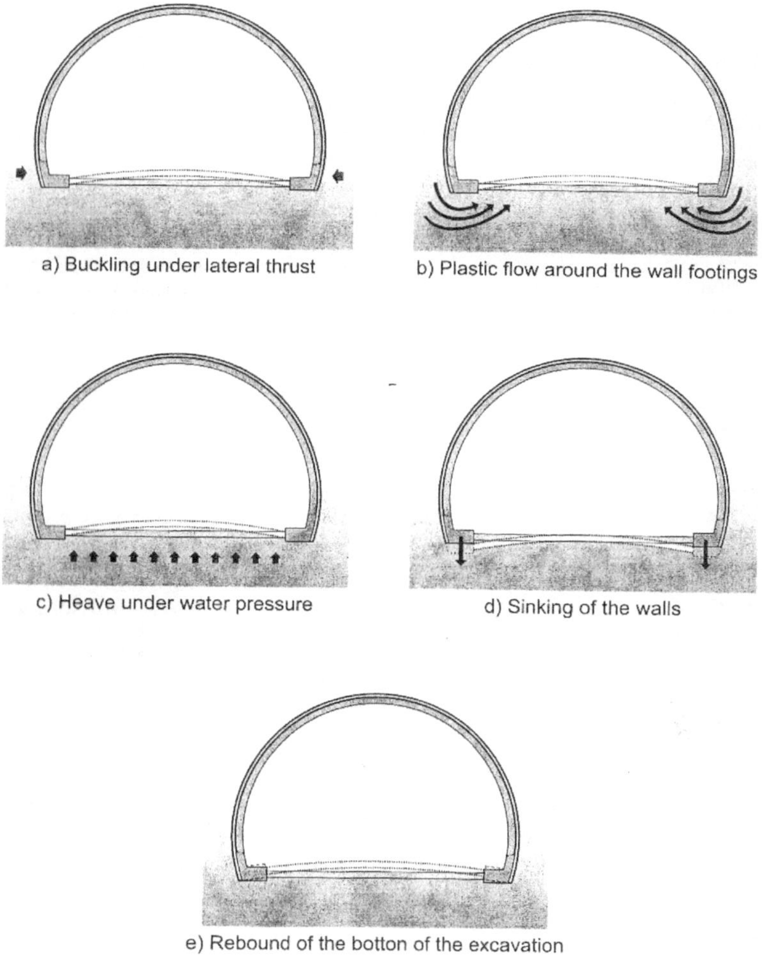

Figure 27. Effects leading to convex deflection of floor slabs.

– Nailing of the lining walls by micropiles in order to add supplementary shear strength against further convergence of the section.

All these operations are very complicated in a completed tunnel and very often require its temporary closure to the traffic.

The El Padrún twin tunnels are a good example of distortion of the floor slab. These 1780 m long tunnels were constructed in 1991 using the NATM technique, with a width of 10 m and a section of 56 m^2. They are located on the Madrid-Oviedo highway, at 12 km from Oviedo. The ground consisted of sandstones, shales, siltstones, and conglomerates of Carboniferous age. Water is present in some faults and at the base of the conglomerates.

Two years after commissioning the floor slabs started to hog, with a longitudinal crack and central heaving, accompanied by some distress at the side culverts. The worst sections corresponded to fault zones or soft water bearing rock. Horizontal convergences of 4–8 mm were measured in the final lining.

A number of interventions were carried out without success such as: drainage of water seepages and water pressures through pipe drains and manholes, repair of pavements, filling of the voids below the floor slab by cement grouting, etc.

After dismissing the possibility of an eventual expansivity of the shales it was concluded that the damages were attributable to a softening of the ground with time under the action of water, thus resulting in increased pressures on the lining and the tunnel invert, with an increase of the deflections by the uplift pressures and some sinking of the walls.

The solutions consisted of

– Demolition of the base slab and the placing of a drainage layer together with a new reinforced concrete slab.
– The same with 2 anchors ϕ 32 mm in each wall at 1.5 m spacing.
– Id. with 3 anchors at 1.1 m centers, together with fracture grouting through 2 m borings below the invert and base of walls. Pipe drains 10 m long (Figure 28).

In the La Encina tunnel the ground softened due to the groundwater affluences and the planar floor slab was unable to strut the walls due to its intersection by the longitudinal drainage ditch.

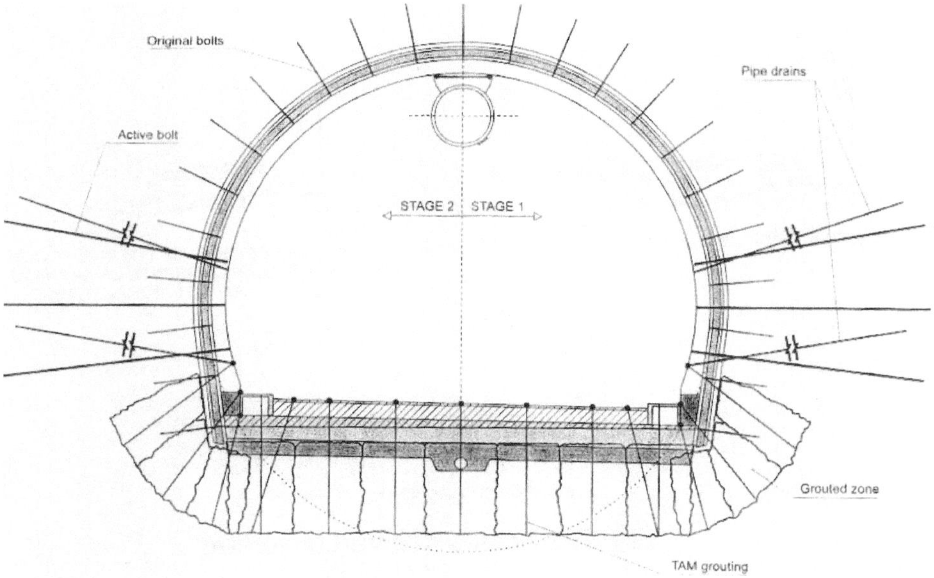

Figure 28. Repair of the El Padrún Tunnel.

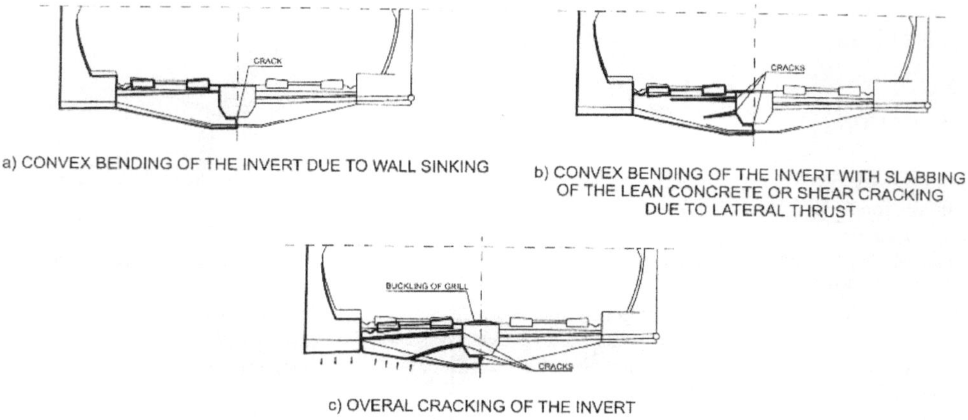

Figure 29. Stages in invert distortion.

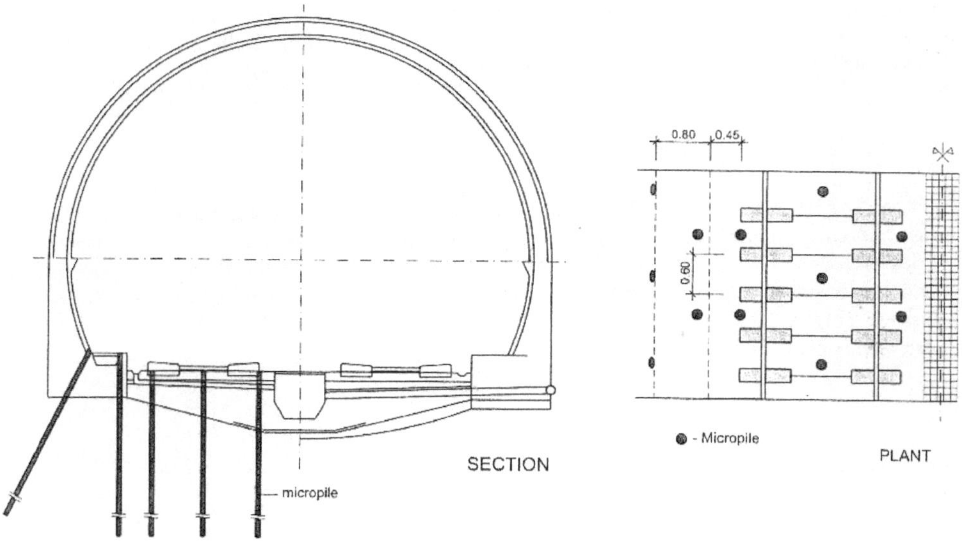

Figure 30. La Encina Tunnel – underpinning solution.

This resulted in bending and shear cracking of the invert (Figure 29). In this case a solution of fixing the base of the walls by means of micropiles acting both as anchorages and as underpinning of the progressively sinking walls was adopted (Figure 30).

ACKNOWLEDGEMENTS

The author is thankful to Mr Diego Gomez, Director of Construction of the Gestor de Infraestructuras Ferroviarias for permitting publication of data referring to the Spanish High Velocity Lines. Thanks are also given to Dr A. Falcon, Head of the Regional Road Section of Asturias for using data from the Fabares tunnel. The information supplied by Prof. Sagaseta,

Mr. Enrique Torres, Dr P. Varona of ITASCA and the companies FCC, GEOCISA, NECSO, and Ferrovial is also acknowledged.

REFERENCES

Aydan, Ö., Akagi, T. and Kawamoto, T. (1993). "The squeezing potential of rock around tunnels: theory and prediction". *Rock Mech. and Rock Eng.*, 2, 137–163.

Barla, G. (2001). "Tunnelling under squeezing rock conditions". *Politecnico di Torino*. 96.

Estefania, S. (2002) "Revestimientos prefabricados en túneles profundos." *Ingeopres n° 110*, November, 16–21 (In Spanish).

Hoek, E. and Marinos, P. (2000). "Predicting tunnel squeezing problems in heterogeneous rock masses such as flysch". *Tunnels & Tunnelling Int.*, 32, Part 1, Nov. 2000, 45–51, Part 2, Dec. 2000, 33–36.

Kirschke, D. (1998). "Sohldruckbegrenzung mittels Knautsch-zone bei Tunneln in schwellendem Gebirge". *Geotechnik 21*, Nr. 3, 192–196.

Kovari, K., Amsatd, C. and Anagnostou, G. (1988). "Design/Construction methods- Tunnelling in swelling rocks". *Proc. 29 U.S. Rock Mech. Symp.*, 17–32.

Kovari, K., Amberg, F. and Ehrbar, H. (2000). "Mastering of Squeezing Rock in the Gothard Base." *World Tunnelling*, June, 234–238.

Rousset, G. (1988). "Comportement Mecanique des Argiles Profondes – Application au Stockage de Dechets Radioactifs". *Thèse Ecole Nationale des Ponts et Chaussées*, Paris.

Singh, B., Jethwa, J.L. and Dube, A.K. (1992). "Correlation between observed support pressure and rock mass quality." *Tunn. & Undergr. Space Techn.*, 7, 59–74.

Steiner, W. (1993). "Swelling Rock in Tunnels: Rock Characterization, Effect of Horizontal Stresses and Construction Procedure". *Int. J. Rock Mech. Min. Sci. & Geomech. Abstr.*, Vol. 30, no. 4, 361–380.

Wittke, M. (2003). Begrenzung der Quelldriicke durch Selbstabdichtung beim Tunnelbau im anhydritfiihrenden Gebirge. WBI Print 13.2 vols. Gluckauf. 112.

Wittke, W. (2000). Stability Analysis for Tunnels. WBI Print 4. Chapter 3. Gliickauf. 415.

Wittke, W. and Pierau, B. (1979). "Fundamentals for the Design and Construction of Tunnels in Swelling Rock." *4th Int. Congress on Rock Mechanics*, Montreux, Vol. 2, 719–729.

Geotechnical Risks in Rock Tunnels – Campos e Matos, Ribeiro e Sousa, Kleberger & Lopes Pinto (eds) © 2006 Taylor & Francis Group, London, ISBN 0 415 40005 8

Geotechnical risk in rock mass characterisation – a concept

I. Poeschl
Salzburg, Austria

J. Kleberger
iC consulenten ZT GmbH, Salzburg, Austria

ABSTRACT: Most rock mass classification schemes fall short in meeting the demands of increasingly complex contractual conditions for today's infrastructure projects, where cost and risk control plays a major role. The concept for rock mass characterisation proposed in this paper systematically aims towards: a) the definition of typical rock mass behaviour and b) the recognition and description of geotechnical risks. The product is the definition of rock mass types (RMTs) that supply the information required for tunnel design as well as for ground-related contractual issues.

1 GEOTECHNICAL RISKS IN TUNNELLING

Geotechnical risks in tunnelling generally stand for hazardous geotechnical conditions that could unfavourably affect a tunnel project and might – in the worst case – cause human fatalities. Less tragic but also significant consequences include damage to equipment, interruption of works, inadequacy of design, contractual claims, etc., all of which eventually lead to delays of project schedule and/or increase of project costs.

Most geotechnical risks can be controlled by adequate design solutions that 1) prevent the occurrence of the "risk event" (e.g. modification of tunnel alignment to avoid intersection of major fault zone), or 2) minimize the consequences of a "risk event" (e.g. adequate design to control deformations and failure or appropriate drainage system for elevated water inflow – Figure 1).

Some geotechnical risks cannot be controlled by economically and/or technically feasible design solutions. The consequences of such a risk event (e.g. catastrophic earthquake) must be assessed and considered in the project risk management plan.

The precondition for risk control and risk management is the timely recognition of potential geotechnical risks for a project. A considerable amount of significant geotechnical risk factors reflects specific rather than standard conditions (e.g. extremely weak materials along fault zones, or karstic fault breccias). Frequently being overlooked in the course of the feasibility study and the design stages, un-foreseen non-standard conditions often give rise to the most significant hazards, damages and contractual disagreements.

Most geotechnical risks in tunnelling are directly related to the properties of the rock mass and to circumstantial influencing factors such as insitu stress, kinematics, groundwater, orientation and dimensions of excavation. The interaction of rock mass properties and influencing factors eventually defines the rock mass behaviour that would be observed during tunnel excavation without application of support/construction measures. Though such a situation does not reflect typical conditions during construction works, the description of rock mass behaviour without support is required for the basic understanding of potential failure modes and risk events.

The scenario actually encountered during and after construction can be referred to as "system behaviour" ÖGG (2001) that relates to deformations and conditions taking place during excavation and after application of routine support measures (Figure 2). A specific rock mass behaviour

Figure 1. Large-scale water inflow, Blisadona Railway Tunnel, Austria, 1999 (Photo: Andreas Knittel).

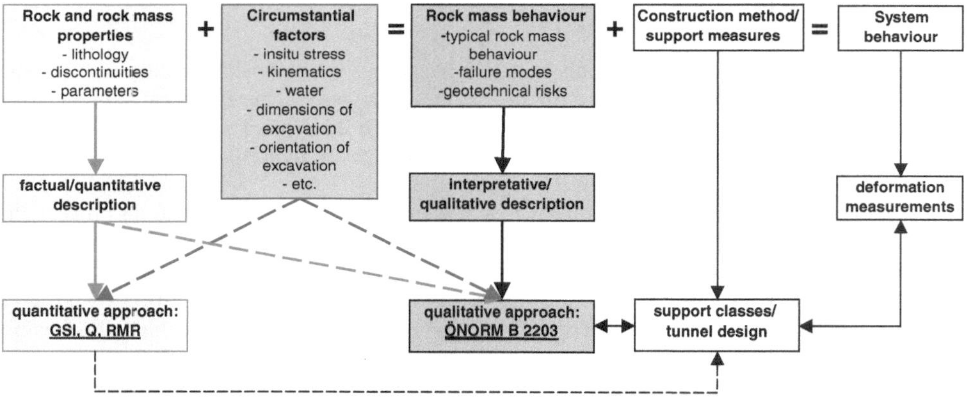

Figure 2. Schematic relations between rock mass properties, rock mass behaviour and support design with quantitative and qualitative classification approaches.

or singular hazardous event that cannot reasonably be controlled by available routine support measures or standard design would lead to unsafe system behaviour, imposing a geotechnical risk.

Ideally, all potential risks related to rock mass characteristics should be covered by a rock mass classification system that provides the conclusive summary of geological/geotechnical conditions expected during tunnelling.

2 QUANTITATIVE AND QUALITATIVE ROCK MASS CLASSIFICATION AND GEOTECHNICAL RISKS

Rock mass properties, including lithological, structural and rock mechanical characteristics and selected circumstantial factors (e.g. groundwater, overburden) are used in standard classification schemes (e.g. GSI, Hoek & Brown (1997) and Hoek, & Marinos (2000), Q, Barton (1988) and RMR, Bieniawski (1989)), that allow for the (semi)quantitative description of typical rock mass materials (see Figure 2). The advantages of such schemes are evidently 1) their international application and acceptance, 2) the relative simplicity and clarity of their application and 3) the reproducibility of classification procedures. Most of these standard schemes were originally developed for jointed rock mass and reach the limits of their applicability in soil-like materials that often prevail in very poor rock mass (e.g. some fault materials). Most quantitative classification schemes lack qualitative descriptions of rock mass behaviour. The consideration of direct interaction between rock mass classes, support classes and deformation measurements during excavation is not foreseen. This may lead either to over-conservative and costly support application, or to the under-estimation of specific hazardous conditions.

In contrast to the quantitative approaches, classification schemes that have been routinely applied in the Eastern Alps (e.g. Austrian standard ÖNORM B 2203 (1994)) were based largely on the qualitative description of rock mass behaviour, Brosch (1986), Laufer (1997), Müller (1978), ÖNORM B 2203 (1994). The advantages include 1) the definition of rock mass classes with typical deformation and behaviour modes 2) The adequacy of the classification and applied support can directly be verified by deformation measurements during tunnel excavation (Figure 2) allowing for a flexible and cost efficient project realisation. The classification scheme proved, 3) to be also suitable in very poor ground where standard description of jointed rock mass reaches its limits of applicability. A safe and efficient performance of rock mass classification according to this qualitative approach, though, requires the profound understanding of geo-mechanical aspects of rock mass behaviour, failure modes and their relation to rock mass characteristics. Because of this prerequisite and the lack of quantitative definitions, the method – though successfully applied in projects worldwide – proofed to be internationally less accepted than quantitative classification schemes.

All of the above described rock mass classification schemes fall short – in one way or the other – in meeting the demands of increasingly complex contractual conditions for today's infrastructure projects, where cost and risk control plays a major role. Inadequate classification schemes are frequently the main cause for contractual disagreements. Adequate ground classification, therefore, is an indispensable tool for tunnel design and a basic part of construction contracts. A classification scheme that allows the assessment of reasonably foreseeable geotechnical risks 1) protects the Client from unjustified claims, and 2) allows the Contractor to request compensation for unforeseeable conditions, both according to international contract regulations or recommendations on risk sharing (e.g. FIDIC (1995), ITA (1988)).

For satisfying these requirements, a rock mass classification scheme must:

a. Involve transparent and reproducible classification procedures based on factual geotechnical data.
b. Provide design parameters that correlate directly with numerical analysis and the design of temporary and final lining.
c. Relate to standard classification schemes, allowing comparative plausibility checks and universal understanding within an international environment of engineering.
d. Describe typical rock mass behaviour and failure modes that relate to routine support classes and can be verified by deformation measurements and observations during tunnel excavation.
e. Describe non-standard hazardous conditions and geotechnical risks that can be controlled by additional non-routine support measures and/or considered in a project specific risk management plan.

New methods and the modification of available standards contributed to the development of rock mass classification throughout the last decade, providing more sophisticated classification tools that take into account contractual as well as technical aspects.

3 PROPOSED CONCEPT FOR ROCK MASS CHARACTERISATION

The concept proposed in this paper combines quantitative and qualitative classification method (Figure 3) and was originally elaborated for project-specific classification systems. A particular

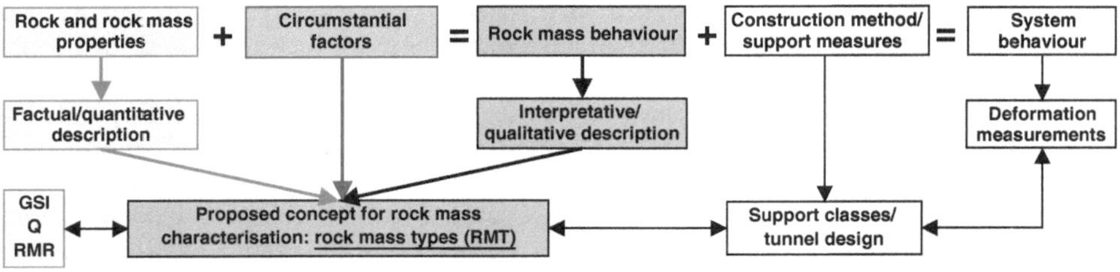

Figure 3. Proposed rock mass characterisation in the frame of ground conditions and tunnel design.

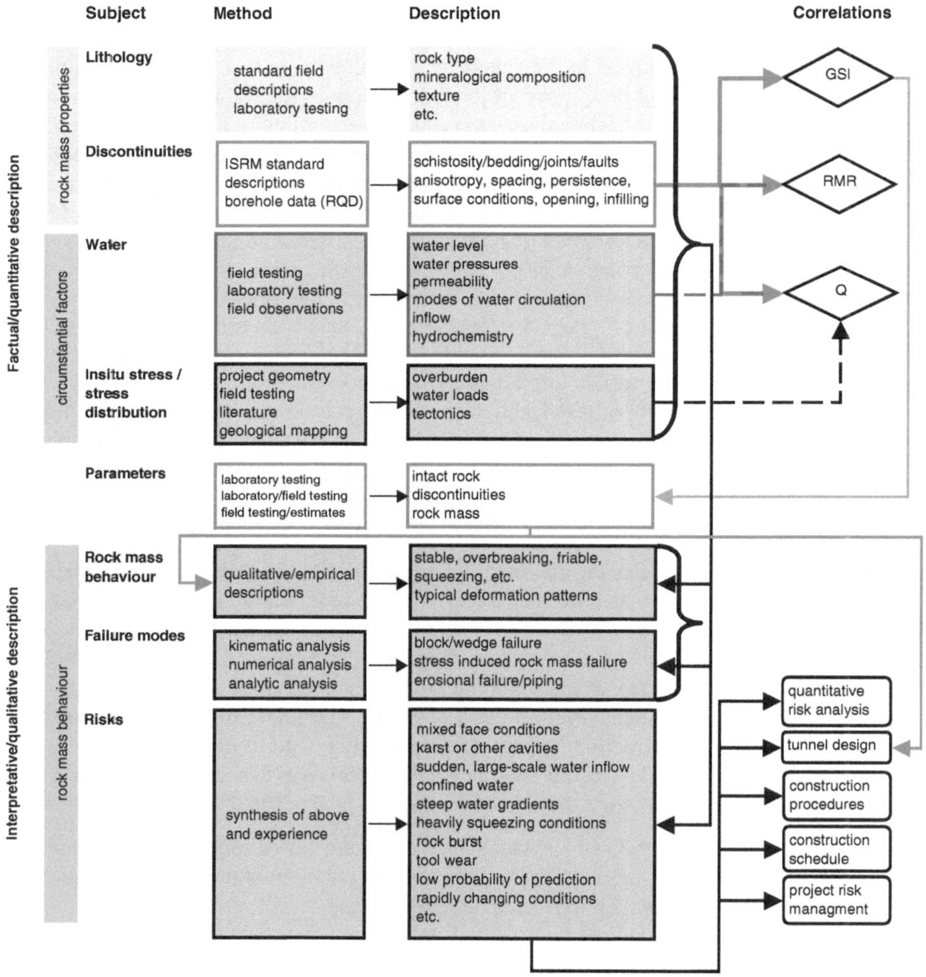

Figure 4. Proposed rock mass characterisation – classification process.

variation of the methodology that supplements the new Austrian Standard for cyclic underground excavation ÖNORM B 2203 (2001) was standardised in official guidelines for rock mass characterisation in ÖGG (2001) and is discussed in Schubert et al. (2001).

The characterisation process permits the definition of "rock mass types" (RMTs) that display typical deformation/failure patterns and/or may impose specific geotechnical hazards. The number of rock mass types individually defined for a project depends on project-specific conditions (such as the complexity of the geology, potential for failure modes and risks). Also because of this inherent and unusual flexibility the concept provides a tool that can handle technical and contractual aspects of non-standard geotechnical conditions with increasing efficiency for all involved parties.

Rock mass descriptions and the assessment of rock mass behaviour, failure modes and risks are based on factual site investigation data, field observations, kinematic, numerical or analytical analysis. Design parameters are derived from statistically evaluated factual data or are estimated according to internationally accepted procedures (e.g. Hoek & Brown criterion, Hoek & Brown (1997), Roclab, Rocscience Inc (2001)). Quantitative standard classifications (e.g. GSI, Q, RMR) are implemented, depending on project specifics.

The process systematically aims towards the definition of typical rock mass behaviour patterns and towards the recognition and description of geotechnical risks. Factual and interpretative, quantitative and qualitative key data are summarised in a single classification sheet for each rock mass type (RMT). Typical rock mass behaviour and geotechnical risks related to non-standard events, issues that are often concealed in text parts of geotechnical reports if mentioned at all, are placed at a prominent position in geotechnical reporting.

For principle relations to rock mass properties and tunnel design refer to Figure 3, for classification process, applied methods and input data to Figure 4.

The geotechnical risks identified and described in the classification sheets should be considered for quantitative risk analysis, development of special design/support/construction measures, construction scheduling and the development of a project risk management plan.

4 CASE HISTORIES

4.1 *Case A – hydropower project – design phase*

Rock mass classification according to the proposed rock mass characterisation scheme was recently performed for a large hydropower project in Turkey (design under progress). The 13 km long power tunnel with 6.6 m OD passes through complex geological conditions in an active tectonic environment. Nine rock types that will be encountered during tunnel excavation (including sedimentary, regional metamorphic, contact-metamorphic, igneous intrusive and volcanic rocks as well as fault materials of all kinds) yielded some 25 rock mass types (RMTs).

Some of these RMTs are fairly similar with regard to basic rock mass parameters (e.g. shear strength, deformation modulus) and typical deformation patterns, but display distinctly different "risk properties". Risks including karst, swelling, excessive tool wear, etc. are directly derived from intact rock and rock mass characteristics. Risk properties including the potential for catastrophic water inflow and for discrete displacements along active faults are interpreted from field observations, literature, seismic data, etc. and require a profound understanding of general geological correlations and engineering geological experience.

Extremely rough terrain and high overburden limited the amount and methods of investigations. This constraint resulted in a more general approach, such as shown in the classification sheets for RMT-7b1, a moderately tectonised carbonaceous conglomerate (Figure 5), and for RMT-11b, clayey fault gouge material (Figure 6).

The numbering system for rock mass types (RMTs) indicates the rock type (lithology) with numbers 1 to 10 (11 stands for fault materials, independently from parent rock), and the degree of tectonisation ("a" for slightly tectonised, "b1" for moderately tectonised, "b2" for highly tectonised).

The power tunnel will be mechanically excavated and the identified risks will essentially influence the selection of a TBM, and the design of the segmental lining.

Rock Mass Type	**RMT-7b1**						
Lithology:	Conglomerate, reddish sandy/silty, calcareous matrix, cemented. Compenents (marble, recrystallised						
Rock mass characteristics:	Thickly bedded to massive, moderately tectonised. Numerous, closely spaced, high-persistence master joints along shear zones. Locally disintegration of matrix materials. RQD typically 50–90%, always > 25% CR typically > 90%, always > 75%						
Discontinuities:							
Bedding:	Anisotropy:	low		Joints:	Sets:	3 main sets	
	Spacing:	600 mm – >2000 mm			Spacing:	<60 mm–600 mm	
	Persistence:	>20 m			Persistence:	<1 m–3 m	
	Conditions:	planar, smooth to rough			Conditions:	planar/stepped, rough to polished	
	Opening:	<5 mm			Opening:	<5 mm to >10 mm	
	Filling:	none			Filling:	rare silt/clay coating	
Parameters:		Intact rock		Rock mass		Discontinuities	
Unit weight, γ	(kN/m3)	24.5–25.5	$n = 4$	24	estimated	-	-
Strength, (UCS)	(MPa)	10–50	$n = 7$	4.58	H&B	-	-
Cohesion, c	(MPa)	-	no data	23	H&B	0	estimated
Friction angle, ϕ	(°):	-	no data	1.22	H&B	25–30	estimated
Young's modulus, E	(MPa):	-	no data	3000	H&B	-	-
Permeability, kf	(m/s):	-	no data	E-05 to E-07	estimated	-	-
CAI	(-)	0.4–0.5	$n = 2$	-	-	-	-
Standard Classification:		GSI:	30–60	RMR:	III (42)	Q:	-
Insitu Stress:	180–430 m overburden. insitu stress exceeding rock mass strength where overburden > 200 m.						
Water:	Dry, where water drained by karstic systems. Potential water head above tunnel unknown. 2–10 l/s per 100 m tunnel length where below groundwater. Locally water inflow from open joints, fractured zones and/or solution cavities with potential for considerable increase during wet seasons. Confined water in relation to low-permeability fault zones possible, with potential for sudden and large-scale water inflow.						
Rock Mass Behaviour / Failure Modes:	Overbreaking to very friable, with rapidly declining deformations. Gravity controlled block / wedge failure, enhanced along sheared discontinuity planes. Stress induced failure in sheared, low-strength rock mass with disintegrated matrix, depending on insitu stress.						
Risks:	Open, karstified masterjoints, forming large blocks in tunnel roof, prone to gravity controlled failure. Sudden and large-scale water inflow from karst cavities / open joints, where tunnel below ground water. Sudden and large-scale water inflow from karst cavities / open joints following the wet season, also along otherwise dry tunnel sections.						
Photo / Sketch:							

Figure 5. Classification sheet for moderately tectonised conglomerate, rock mass type RMT-7b1.

Rock Mass Type	RMT-11b						
Lithology:	Fault gouge, (kakirite, cataclasite), rock fragments of various dimensions in clayey/silty matrix, matrix supported						
Rock mass characteristics:	Completely sheared, disintegrated and crushed rock mass, slickensided throughout. RQD = 0% (always) CR <25% (always)						
Discontinuities:							
Bedding:	Anisotropy:	-		Joints:	Sets:	-	
	Spacing:	-			Spacing:	-	
	Persistence:	-			Persistence:	-	
	Conditions:	-			Conditions:	-	
	Opening:	-			Opening:	-	
	Filling:	-			Filling:	-	
Parameters:		Intact rock		Rock mass		Discontinuities (fault planes)	
Unit weight, γ	(kN/m3)	-	-	20	estimated	-	-
Strength, (UCS)	(MPa)	-	-	0,26	estimated	-	-
Cohesion, c	(MPa)	-	-	15	$n=5$	0	estimated
Friction angle, ϕ	(°):	-	-	0,1	$n=5$	10–15	estimated
Young's modulus, E	(MPa):	-	-	100	estimated	-	-
Permeability, kf	(m/s):	-	-	<E-07	estimated	-	-
CAI	(-)	-	-	-	-	-	-
Standard Classification:		GSI:	not applicable	RMR:	not applicable	Q:	not applicable
In situ Stress:	Variable overburden. Insitu stress considerably exceeding rock mass strength. Heterogeneous and asymmetric stress distribution possible, where large competent shear bodies embedded in fault gouge matrix.						
Water:	Acting as groundwater barrier, confining water circulating along fractured zones and/or karstic systems.						
Rock Mass Behaviour/Failure Modes:	Squeezing conditions, with pronounced, slowly declining deformations. Stress induced rock mass failure.						
Risks:	Heavily squeezing conditions with large, slowly declining or continuous deformations. Swelling of clay minerals in fault gouge materials. High water pressures, confined water and steep gradients along interfaces of jointed rock rock mass and fault gouge materials. Stickiness of clay materials. Mixed-face conditions where competent shear bodies embedded in fault gouge materials. Discreet displacements along active faults possible.						
Photo/Sketch:							

Figure 6. Classification sheet for fault gouge material, rock mass type RMT-11b.

4.2 *Case B — urban infrastructure tunnel – tender design*

A similar approach was applied for the tender design of a shallow tunnel passing through schists and granites that are weathered and disintegrated to various degrees. The rocks occur in

Figure 7. Typical failure in highly sheared schists with granitic dykes, rock mass type X3 (left); and erosional failure in disintegrated granitic dyke (right).

close association along an intrusive interface overprinted by intense tectonic shearing. In contrast to the case history in Turkey, the alignment corridor, with maximum overburden reaching 30 m, could be investigated thoroughly. Six rock mass types (referred to as geotechnical units) were defined.

Typical rock mass characteristics/behaviour and failure modes for schists and granites could be observed in cut-slopes of an adjacent construction pit (see Figure 7). Typical behaviour patterns and risks for granitic rock mass have been assessed for previous infrastructure projects during the construction of some of these projects.

With the drive direction already known, detailed kinematic failure analysis could be performed and modes of wedge failure are shown in schematic sketches for each rock mass type. The most relevant geotechnical properties, circumstantial conditions and rock mass behaviour modes were summarised in classification sheets, as shown in Figure 8 for geotechnical unit X3, where sheared schists are intersected by disintegrated granitic dykes.

Conventional cyclic tunnel excavation is foreseen for the project. The rock mass characterisation allowed the development of routine temporary support classes as well as special measures to be employed in case of "risk events" (e.g. piping and erosional failure along disintegrated granitic dykes, etc.).

5 SUMMARY AND CONCLUSIONS

The concept for rock mass characterisation proposed in this paper combines quantitative and qualitative rock mass classification systems. The process can be individually adjusted by employing a variety of methods. Key elements including rock and rock mass properties, circumstantial influencing factors, geomechanical parameters, typical rock mass behaviour, failure modes and geotechnical risks are described in a classification sheet for individual rock mass types that are correlated to quantitative standard classification systems. All descriptions and assessments are performed according to international and/or national standards/methods, allowing for a transparent and reproducible classification process. The scheme systematically aims towards a) the definition of typical rock mass behaviour and b) the recognition and description of geotechnical risks. The product is the definition of rock mass types (RMTs) that supply the conclusive information required for tunnel design as well as for ground-related contractual issues.

Geotechnical Unit:		**X3**			
Lithology:		Mica schists, sericitic schists, migmatitic gnaissic schists intersected by granitic dykes. Weathering grades W4, W5 prevailing.			
Discontinuities:					
Schistosity:	Anisotropy:	high	Joints:	Sets:	2 main sets
	Spacing:	<60 mm		Spacing:	<60 mm–600 mm
	Persistence:	10–20		Persistence:	<1 m–3 m
	Conditions:	undulating, polished		Conditions:	planar/stepped, planar, locally polished
	Opening:	<5 mm		Opening:	<5 mm
	Filling:	soapy surfaces, frequently plastic clay		Filling:	silt/clay coating, locally plastic
Intact Rock Parameters:					
	Density (g/cm³):	2.0–2.4			
	UCS (MPa):	<5			
	T (MPa):	<1.5			
	Abrasivity:	heterogeneous, typically low	estimated, no data		
Discontinuity Parameters:					
	c (MPa):	0	estimated		
	ϕ (°):	20	estimated		
Standard Classification:					
	GSI:	10–20			
	RMR:	V			
Rock Mass Parameter:		Refer to report.			
Rock Mass Characteristics:		Highly anisotropic rock mass, highly tectonised.			
Schematic Sketch:					

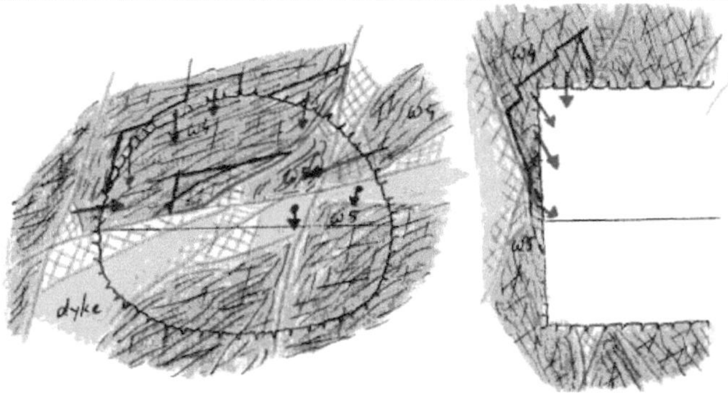

Insitu Stress:	Insitu stress usually exceeds overall rock mass strength.
Water:	Joint water. Damp to wet, dripping to flowing from granitic dykes. Water inflow typically below 5 l/s, rarely up to 15 l/s per 100 m. Local water inflow from pegmatitic dykes, fractured zones and/or interface jointe/soil-like rock mass may reach 10l/s to 20 l/s. Water within granitic dykes likely to be confined and subject to high water pressures.
Rock Mass Behaviour:	Very friable, rapidly ravelling, squeezing (depending on overburden). Relaxation sliding along low-strength schistosity planes. Stress induced failures of low-strength rock mass, enhanced by heterogeneous stress distribution related to shear planes and faults.
Remarks/Risks:	Erosion/stress induced failures enhanced in association to granitic dykes (W5/W4). Mixed face conditions.

Figure 8. Classification sheet for highly tectonised schist with granitic dykes, rock mass type X3.

ACKNOWLEDGEMENTS

The authors thank the project's Clients for the permission to publish project specific data of the case histories discussed. Further, reference is made to all parties and persons – too many to be named – that contributed to the concept of rock mass characterisation presented in this paper.

REFERENCES

Barton, N.R. (1988). Rock Mass Classification and Tunnel Reinforcement Selection using the Q-system. – *Proc. Symp. Rock Class. Eng. Purp.*, ASTM Special Technical Publication 984. Philadelphia. pp. 59–88.

Bieniawski, Z.T. (1989). Engineering Rock Mass Classifications. – Wiley.

Brosch, F.J. (1986). Geology and Classification of Rock Masses – Examples from Austrian Tunnels. – *Bull.Int.Assoc.Eng.Geol.*, no. 33, pp. 31–37.

FIDIC. (1995). Conditions of Contract for Design-Build and Turnkey, Parts I and II, First Edition.

Hoek, E. & Brown, E.T. (1997). Practical Estimates of Rock Mass Strength. – *Int. J. Rock. Mech. & Mining Sci. & Geomec. Abstr.* 34(8), 1165–1186.

Hoek, E. (1999). A discussion on acceptable criteria for temporary support and final linings of large span transportation tunnels in poor rock. – Vancouver.

Hoek, E. & Marinos, P. (2000). Deformation. Estimating Rock Mass Strength. Predicting Squeeze. – *Tunnels & Tunelling International.*

ITA. 1988 to 1992. ITA Recommendations on Contractual Sharing of Risks. – *Tunneling and Underground Space Technology*, Vols 3, 5 and 7.

Laufer, H. (1997). Rock Classification Methods Based on Excavation Response. – *Felsbau* 15, no. 3.

Müller, L. (1978). Removing Misconceptions on the New Austrian Tunnelling Method. – *Tunnels Tunnelling* 10, pp. 29–32.

ÖGG. (2001). Richtlinie für die Geomechanische Planung von Untertagebauarbeiten mit zyklischem Vortrieb.

ÖNORM B 2203. (1994). Untertagebauarbeiten, Werkvertragnorm, *Entwurf*, 1.

ÖNORM B 2003–1, Ausgabe 2001–12–01. Untertagebauarbeiten – Werksvertragsnorm, Teil 1: Zyklischer Vortrieb.

Rabcewicz, L. (1964). The New Austrian Tunnelling Method. – *Water Power*, pp. 453–457.

Rocscience Inc. 2001. RocLab.

Schubert, W., Goricki, A., Button E., Riedmüller, G., Pölsler, P., Steindorfer, A. and Vanek, R. (2001). Consistent Excavation and Support Determination for the Design and Construction of Tunnels. – *Felsbau*, 5/2001.

Geotechnical Risks in Rock Tunnels – Campos e Matos, Ribeiro e Sousa, Kleberger & Lopes Pinto (eds) © 2006 Taylor & Francis Group, London, ISBN 0 415 40005 8

Risk control at the design of a 13 km long railway tunnel in Austria

J. Daller
iC consulenten Ziviltechniker GesmbH Vienna, Austria

ABSTRACT: Risk assessment and risk control are very important aspects from the very beginning of a design process to the successful implementation of large scale tunnel projects. The risk management used for the tender design of the Wienerwald Tunnel Project is briefly presented.

1 INTRODUCTION

The more than 13 km long Wienerwald Railway Tunnel Project comprises two single track tunnels, each about 10 900 m long, a 2236 m long double track section with a 409 m long enlarged cross section at the transition to the twin tubes, a ventilation cavern with a permanent ventilation shaft, three optional temporary ventilation shafts during construction, three permanent emergency exits, 22 cross passages connecting the single track tubes at every 500 m distance, inclined mucking galleries equipped with conveyor belts and a temporary construction access (see also Figure 2). For the tender of the project 3 construction lots were combined (Lot WT2, Lot LT26, and Lot TF3). The geological conditions along the tunnel are characterized by Molasse and Flysch formations. The tunnel will cross several extensive fault zones. The overburden ranges from about 10 to 200 m.

2 DESIGN ASPECTS FOR RAILWAY TUNNELS

The design of railway tunnels in Austria is generally based on our national standard for the design of "high capacity railway tunnels", HL-AG (2002).

Designing modern high speed railway tunnels special attention has to be paid to following aspects:

- Aerodynamics
- Safety facilities and emergency exits
- Fire hazards
- Track work design including vibration control
- Tunnel lining

2.1 *Aerodynamic requirements*

According to UCI-Code 779–11 the minimum cross section area of a tunnel profile has to be calculated during the preliminary design phase considering aerodynamic effects. In Austria the following permissible pressure changes are specified for common (not pressure tight) railway carriages:

- Double track tunnels: 4000 Pa/4s;
- Single track tunnels: 2500 Pa/4s.

For a design speed of 200 km/h (maximum speed = 250 km/h), a tunnel length of 10 km and using a concrete track slab the following minimum inner cross section area is required:

- 44.2 m2 for a single track tunnel, and
- 76.5 m2 for a double track tunnel.

In double track tunnels the distance between the track axes is 4.70 m. However, reducing the spacing between the tracks would not result in a smaller tunnel profile, since the required cross section areas are controlled by aerodynamic aspects. For most of the railway tunnels (single and double track) in Austria aerodynamic requirements control the size of the tunnel cross section.

2.2 *Safety facilities*

For tunnels longer than 500 m an emergency escape system must be designed. The maximum spacing of escape facilities is specified with 500 m.

In a single tube tunnel either vertical emergency shafts or horizontal (gradient < 10%) emergency galleries (clearance profile 2.25 × 2.25 m) must be provided for. If emergency galleries are longer than 150 m they must be enlarged (clearance profile 3.60 × 3.50 m plus walk-way 1.20 × 2.20 m) so that ambulance cars and small fire brigade cars can enter the galleries.

In parallel single track tunnels cross passages between the tubes serve as emergency escape routes. The cross passages must be closed at either sides with doors. The space between the doors must be at least 12 m long. The ventilation fans installed in each cross passage and emergency exit must have the capacity to keep the escape routes free of smoke in case of a tunnel fire.

For very long tunnels (exceeding 20 km) underground emergency stop facilities have to be designed.

2.3 *Fire hazards*

a) *Fire fighting facilities*

For fire fighting in case of a tunnel fire each tunnel tube must be equipped with one water main. The water mains are normally placed in a cable duct or in case of a double track tube between the tracks. Hydrants are located at the tunnel walls or in niches with maximum spacing of 150 m. The water pressure at the hydrants is specified with 6–10 bars.

b) *Structural design*

Tunnel structures shall be designed and dimensioned that in case of a tunnel fire the damage of the tunnel is limited, third parties (e.g. user of a road above the railway tunnel) will not be affected and the period of a tunnel closure for repair will be the shortest possible.
The design targets shall include the following topics:

- maintain stability of the structure for a defined period, in order to allow passengers to escape and to evacuate buildings or infrastructures on top of the tunnel;
- limit deformations of the structure and surface settlements;
- avoid tunnel collapse if this affects surface structures;
- maintain water tightness for submersed tunnel; and
- enable the possibility for repair considering costs and time.

c) *Design fire*

In several European countries design fires (temperature curves versus time) are defined. In the revised Austrian standard a temperature versus time curve for mixed traffic (passenger trains and cargo trains are operated on the respective railway line) called "EBM" will be specified (see Figure 1). The duration of the design fire considered for the structural design of the tunnel lining is influenced by several factors (see protection levels and classification matrix in Table 1).

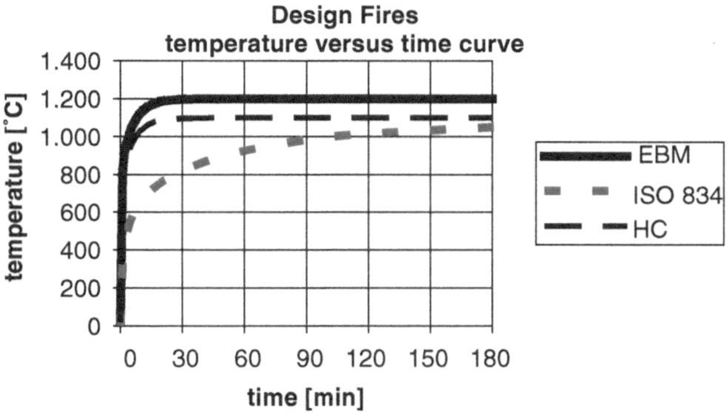

Figure 1. Design fire acc. to proposed Austrian guideline.

Table 1. Classification matrix for design fire.

Protection level	Possibility for repair	Mined tunnels longer than 200 m Diversion of traffic not possible	Diversion traffic possible level
0	Type A, B	No design required	No design required
1	Type A	EBM × min *	EBM × min *
	Type B	EBM 60 min	EBM × min *
2	Type A	EBM 90 min	EBM 60 min
	Type B	EBM 120 min	EBM 90 min
3		EBM 180 min	EBM 150 min
Special cases		Special design considerations	

*, Duration of design fire as required for evacuation of buildings and infrastructures above the tunnel (time needed for intervention at the ground surface above the tunnel).

d) *Protection levels*

Four protection levels are defined:

- Protection level 0: no effect on surface structures or third parties.
- Protection level 1: only minor structures or infrastructure on the surface might be affected.
- Protection level 2: major structures (e.g. residential buildings) and infrastructures (e.g. main roads, railway lines etc.) above the tunnel might be affected.
- Protection level 3: major buildings with difficult evacuation (e.g. hospitals, schools, airports, etc.) and important infrastructures (e.g. main water supply line) above the tunnel might be affected; or submersed tunnels (e.g. river crossing, tunnel under ground water table etc.).

e) *Possibility for repair*

This classification considers the possibility for repair or reconstruction of an affected tunnel with respect to necessary time, repair costs and accessibility.

- Type A: easy repair/reconstruction
- Type B: difficult repair/reconstruction

f) *Specification of design fire for structural design of tunnel linings*

The design fire considered for the structural design of tunnel linings shall correspond with the levels as specified in the table below.

g) *Measures for structural fire protection*

Without any special protection or measures un-reinforced or reinforced tunnel lining can withstand temperatures beyond 1000° C only a very limited time until progressive spalling of the concrete starts.

In the recent year extensive research and testing programmes were carried out to find technical sound and economic solutions for fire protection and avoidance of spalling.

Basically three different solutions were investigated:

- Application of fire protection panels or layers of protective material.
- Increasing the concrete cover for the inner layer of the structural reinforcement and to add an additional layer of net reinforcement to avoid spalling.
- Adding polypropylene fibres to the concrete mix (can also be applied with un-reinforced lining).

The test results showed that adding polypropylene fibres is the most effective and also economic solution to avoid spalling and keep the temperature of the reinforcing steel an concrete at an acceptable temperature.

Special attention has to be paid to the workability (pumping, compacting) of the concrete by adding polypropylene fibres (about 1.8–2.0 kg/m^3 concrete).

2.4 *Tunnel lining, water proofing and drainage system*

Especially using a shielded tunnel boring machine (TBM) with a segmental lining the designer and client (owner) have to decide whether to apply a single or double shell lining. This decision has to be made in a close context with the drainage and water proofing system.

For the Wienerwald tunnel project a risk analysis and cost comparison was made for selecting the proper lining in case of TBM – drive. The analysis showed clearly that a double shell lining consisting of prefabricated concrete segments and an *insitu* inner lining is the best solution. A water proofing membrane will be installed between outer and inner lining and the mountain water will be drained off by longitudinal drainage pipes at both side walls.

3 WIENERWALD TUNNEL PROJECT

3.1 *General*

The project was tendered in autumn 2003. For the tender three construction lots were combined and tendered together:

- Lainzer Tunnel Lot LT26: double track tunnel, 1.717 m long.
- Wienerwald Tunnel Lot WT2: double track tunnel, 110 m long;
enlargement of double track tunnel, 409 m long; and
single tube tunnels, each about 10 900 m long.
- Tullnerfeld Lot TF3: earth works (embankment for railway line, noise protection walls) using tunnel excavation material from about 2 times 5 km single track tubes.

The project was tendered in two design alternatives. It was the first time in Austria that full sets of tender documents were prepared for different construction methods. For the single track tunnels from the western portal to the emergency ventilation cavern design drawings and tender documents were prepared for using either TBM's or applying the NATM.

In order to be able to compare and evaluate the bids for the different construction methods a risk assessment was carried out for both design alternatives. The risk assessment considered geotechnical conditions impairing tunnel stability or construction progress, possible damage to the tunnel lining, problems occurring with equipment during tunnel driving and impact on environment and surface structures.

Acceptable, potential remaining risks, which could not be eliminated during the design process, were quantified and will be considered in the bid evaluation.

Risk sharing between the contractor and the client is clearly stipulated in the tender documents. For bid evaluation only remaining risks for which the client is responsible will be taken into account.

Submission date for the construction tenders was end of February this year. Only four bidders submitted their offers. The bidding companies come from Austria and Germany, with sub-contractors from Sweden and Switzerland. One bidder submitted offers for both alternatives, the other three groups submitted only alternative B with the application of TBM's. At the moment the offers are being checked and evaluated.

3.2 *Risk management during tender design*

The risk assessment process during the tender design is shown on the flow chart in Figure 3.

For conventional tunnel excavation the following risk parameters were identified:

a) Due to rock mass conditions (tunnel stability, excavation equipment, tunnel lining) such as

- rock pressure and tunnel deformation,
- loosening of rock mass under shallow cover,
- swelling pressure,
- occurrence of methane gas,
- instability of tunnel face and excavated rock surface,
- over-break and smoothness of excavation profile,
- inrush of mountain water (temporary and permanently),
- frequently changing rock conditions,
- sticking ground conditions when encountering fault material or clayey material, and
- abrasiveness of rock mass with high quartz content.

b) Risks related to environmental aspects such as

- surface settlements affecting buildings and infrastructures,
- effecting springs and wells,
- construction vibrations, noise and dust, and
- train vibrations during operation.

For tunnel construction using a TBM the following risk parameters were identified:

a) Due to rock mass conditions (tunnel stability, excavation equipment, tunnel lining) such as

- rock pressure and tunnel deformation,
- getting stuck in major fault zones,
- loosening of rock mass under shallow cover,
- swelling pressure,
- occurrence of methane gas,
- instability of tunnel face, over-break in front or above cutter head,
- over-break in the area of the shield,
- encountering big blocks embedded in soft rock matrix,
- inrush of mountain water (temporary and permanently),
- sticking ground conditions when encountering fault material or clayey material, (clogging of cutter disks, opening in the cutter head, conveyor belts),
- abrasivity of rock mass with high quartz content, and
- deformation of tail shield due to excessive rock pressure in major fault zones.

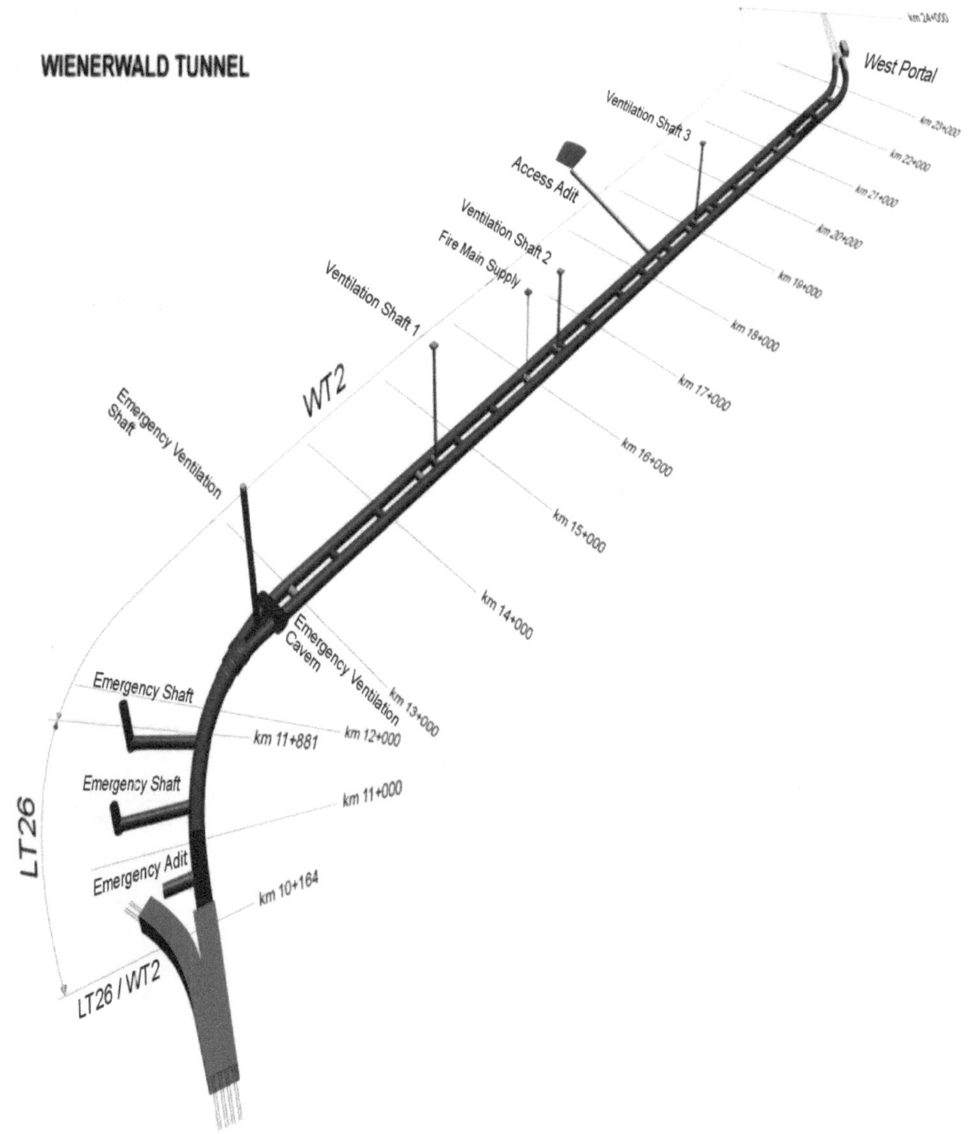

Figure 2. Schematic layout of Wienerwald tunnel – construction Lots WT2 and LT 26.

b) risks related to equipment such as

- failure of main bearing,
- fire hazard on the TBM, and
- break-down of mucking system (conveyor belts) and interrupting of material supply.

c) risks related to tunnel lining,

- insufficient bedding of segmental lining,
- increase of loads onto segmental lining due to close spacing of parallel tubes, and
- increase of loads onto segmental lining due to excavation of cross passages.

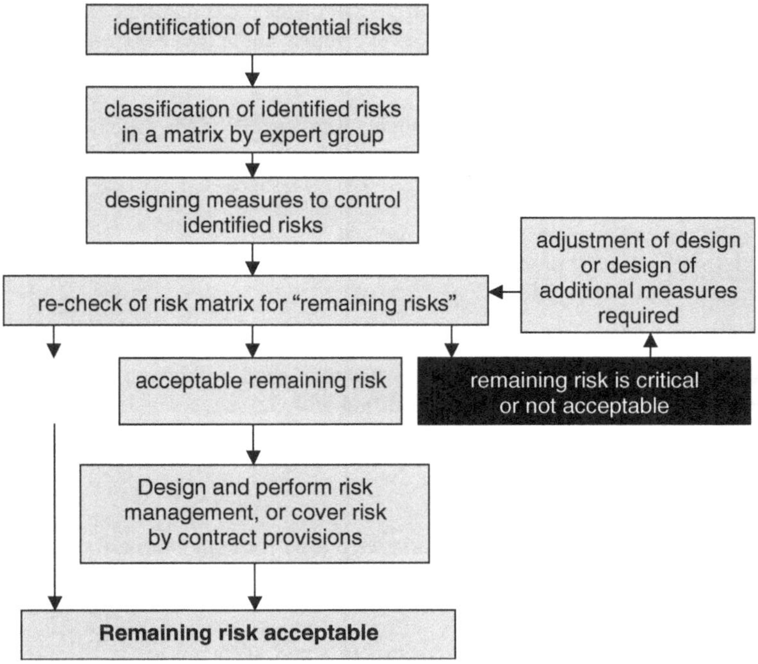

Figure 3. Flow chart for risk assessment during design.

d) risks related to environmental aspects such as

- surface settlements affecting buildings and infrastructures,
- effecting springs and wells,
- construction vibrations, noise and dust, and
- train vibrations during operation.

Most of the risk parameters identified could either be eliminated during the design, or considered in the bill of quantities or covered by provisions in the contract documents. The remaining risks, which could not be eliminated or fully covered by the tender design, were quantified for each construction method. The quantification was used to make both alternatives comparable in the tender evaluation. In the quantification only risks were considered which belong to the client.

The remaining risks quantified for conventional tunneling were:

- surface settlements in sections with low overburden affecting buildings and infrastructures;
- effecting springs and wells due to lowering of mountain water table; and
- construction vibrations in sections with low overburden and hard rock conditions.

The remaining risks quantified for using a TBM were:

- getting stuck in major fault zones;
- loosening of rock mass under shallow cover;
- instability of tunnel face, over-break in front or above cutter head;
- deformation of tail shield due to excessive rock pressure in major fault zones;
- insufficient bedding of segmental lining;
- surface settlements in sections with low overburden affecting buildings and infrastructures;
- effecting springs and wells due to lowering of mountain water table; and
- construction vibrations in sections with low overburden and hard rock conditions.

The most challenging situation for the TBM-drive will be the crossing of the major tectonic faults. Fortunately the faults will be encountered in a very obtuse angle. Even so, one of the main faults ("Rehgraben Fault") has an extension of about 80 m and the tunnel passes under an overburden of nearly 200 m.

In the tender documents minimum requirements were specified for the TBMs, such as:

- minimum torque of cutter head (for the range of 0 – 2 r/min),
- minimum torque of cutter head for start up (for at least 60 s),
- minimum total thrust per meter of shield length,
- minimum design load (rock pressure) for tail shield, and
- minimum range of overexcavation.

In addition the TBM must be equipped to install an inclined pipe roof umbrella and inclined rock dowels through the shield as close as possible to the cutter head (for the rock dowels the maximum distance from the front end of the shield is specified with 3.5 m).

3.3 *Tender evaluation*

For evaluating the best bid several other criteria were used for both alternatives beside the tender price. The so-called "modified offer" consists of:

- tender price,
- increase or reduction due to the offered construction period (factor a),
- increase or reduction considering the remaining risk (factor b), and
- additional costs for following works (factor c).

For both of the tender alternatives (conventional tunneling and TBM-drive) an average construction period was estimated by the designer. In addition the client calculated the costs for his services (e.g. project management, construction supervision, geological documentation, interpretation of geotechnical monitoring, etc.) for the estimated construction period and also the costs incurred for each day shortening or extension. In his offer every bidder had to calculate the overall construction period according to his progress rates for the projected underground conditions. If the contractor cannot meet the offered construction period he has to pay penalty.

Factor (a) includes the costs of services calculated by the client and the increase or reduction according to the bidders overall construction period.

Factor (b) considers the quantified remaining risks for each alternative. For the TBM alternative also the technical data of the offered machine compared to specified values were included in factor (b).

Factor (c) takes into consideration different costs between the two alternatives which will be incurred during subsequent construction works (e.g. track works).

Finally, the best offer is the bid with the lowest "modified offer", provided the bidder meets all the other criteria and preconditions (e.g. reference projects, financials specified in the design documents.

4 CONCLUSION

Risk control became an essential tool for the design of tunnel projects. The risks can already be minimized in the design process once they are identified and assessed correctly. Especially in case a project is tendered with different construction methods a risk analysis will help to make offers comparable.

REFERENCES

HL-AG, ÖBB, BEG. 2002. Richtlinie für das Entwerfen von Bahnanlagen – Hochleistungsstrecken.

Lemmerer J., Kusterle W. and Pehersdorfer H. 2002. Brandschutz im Tunnelbau, Betontag.

PGLT. 2003. Risikoanalyse Wienerwaldtunnel.

Vavrowsky G.M., Ayaydin N. and Schubert, P. 2001. Geotechnisches Sicherheitsmanagement im oberflächennahen Tunnelbau, Felsbau 19. Nr.5.

Geotechnical Risks in Rock Tunnels – Campos e Matos, Ribeiro e Sousa, Kleberger & Lopes Pinto (eds) © 2006 Taylor & Francis Group, London, ISBN 0 415 40005 8

Koralm tunnel – benefits of a structured design and investigation process – the client's view

G. Harer
ÖBB/Austrian Federal Railways, Austria

ABSTRACT: Ever since Austria became a member of the European Community, we have had to deal with an enormous increase in traffic. Also for the future, in particular because of the entry of several south-eastern European countries to the EC, a significant traffic increase is predicted. In this task, Austria's government defined, in conformity with European intentions, main railway axes to be improved for future requirements. Investments of almost 30 billion Euros in total are planned for the future improvement of Austria's railway infrastructure within the next decades. One of the key railway lines in the TEN is the so-called Pontebbana corridor. An essential part of this Pontebbana axis is the Koralm tunnel with a planned length of 32.8 km. It will underpass the Koralpe, a stretch of mountains at the border of the provinces of Carinthia and Styria. Around 30% of the tunnel will encounter tertiary sediments (clayey to silty sands) with an overburden of less than 200 m. 70% of the tunnel will pass through crystalline rockmass with a maximum overburden of up to 1.200 m.

1 INTRODUCTION

Due to Austria's location in the centre of Europe, within the Alps, Austria has played an important role for European passenger and goods traffic since historic times. Crossing the Alps has historically been one of the biggest challenges in the construction and maintenance of traffic routes. The construction of tunnels was one of the solutions to deal with the topographical problems.

The construction of extensive railway routes during the last decades of the Austrian – Hungarian Monarchy in particular and the major motorway projects in the sixties, seventies and eighties of the 20th century resulted in more than 300 road tunnels and around one hundred km of railway tunnels in Austria.

2 THE AUSTRIAN HIGH CAPACITY RAILWAY NETWORK

Ever since Austria became a member of the European Community, Austria have had to deal with an enormous increase in traffic. Also for the future, in particular because of the entry of several south-eastern European countries to the EC in May, 2004, a significant traffic increase, with an above-average increase of road transport, is predicted.

However, the changed political and environmental situation in Europe in the future will require a transport system that is suited to the dimensions of the European Continent. Most notably the high speed traveling by rail is an environmentally compatible solution whilst at the same time enabling long-term sustainable mobility. The railway companies are currently working hard to create a trans-European high-speed railway network as part of a Trans European Network (TEN). In this task, they are receiving political support at both national and international level.

Austria's government defined, in conformity with European intentions, five main railway axes to be improved:

- the Danube axis from Eastern Europe via Vienna to Salzburg and Germany
- the Pontebbana axis from Eastern Europe via Vienna to Italy and the Mediterranean Sea
- the Pyhrn axis from Germany via the central regions of Austria to the former Yugoslavian countries
- the Tauern axis from Germany via Salzburg to Italy and Slovenia
- the Brenner axis to Germany and Italy as well as Switzerland.

Investments of almost 30 billion Euro in total, about 1 billion Euro the year, are foreseen for the future improvement of Austria's railway infrastructure within the next decades. Meanwhile in Austria around 200 km of new railway tunnels are being planned or are under construction, some are actually in operation.

Considering already the number and the extension of the projects, in particular of future subsurface structures, special efforts have to be undertaken to guarantee an efficient process to keep time and cost estimated. Some aspects of a structured design and investigation process of the Koralm tunnel will be presented below.

3 KORALM TUNNEL

One of the key railway connecting lines in the TEN is the so-called Pontebbana corridor. It represents the easternmost crossing of the Alps and links eastern Europe, Vienna, southern Austria and northern Italy (see Figure 1). Main efforts to improve this connection were implemented in several stretches along the line. One of these projects is the Koralm railway in the South of Austria. This new stretch will have a total length of approximately 130 km, and will underpass the Koralpe, mountain range between the provinces of Carinthia and Styria. It will decrease the travel time between the provincial capitals of Graz and Klagenfurt from the present time of three hours to one hour, Vavrovsky et al. (2001). The most prominent tunnel along this stretch will be the Koralm tunnel. This double tube tunnel will have an length of approximately 32.8 km, making it Austria's longest tunnel and the seventh longest tunnel project in the world (http://home.no.net/lotsberg/). The maximum overburden will reach almost 1 200 m.

4 GEOLOGICAL OUTLINE

The mountain range which will be crossed by the Koralm tunnel consists of a polymetamorphic crystalline basement. Predominant lithology consists of mylonitic gneisses and micaschists, with occasional marbles, amphibolites and eclogites. The crystalline basement is bounded by master faults which have generated tertiary basins on both sides of the mountain range. These occur as the Weststeirische Becken in the East, and the Lavantaler Becken in the West. The sediments of both Tertiary basins encountered by the tunnel are mainly clastic deposits of fluviatile and marine origin (see Figure 2).

The recent morphological features of the Koralm massif were formed by Tertiary to Quaternary brittle faulting, weathering and erosion. Residual soils, generated by deep reaching in situ weathering and periglacial debris, cover the bedrock.

5 PROJECT STATUS

HL–AG was authorised by the Austrian government in 1995 to undertake the planning and the design of the Koralm railway including the Koralm tunnel. Meanwhile the route assessment, the tunnel system decision and the environmental impact assessment could be concluded. In a next

Figure 1. The Pontebbaba axis.

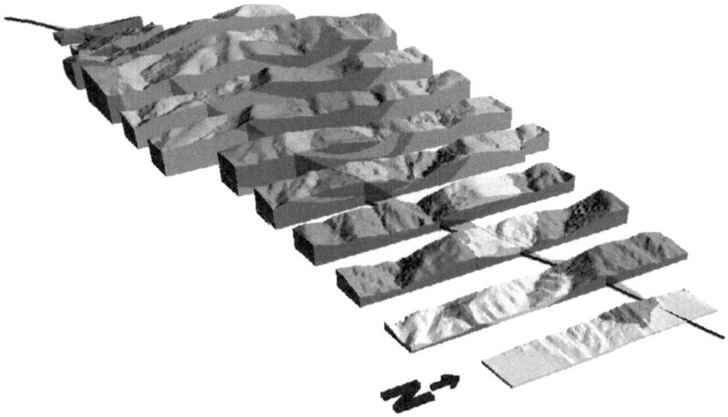

Figure 2. Geological overview of the Koralm Tunnel.

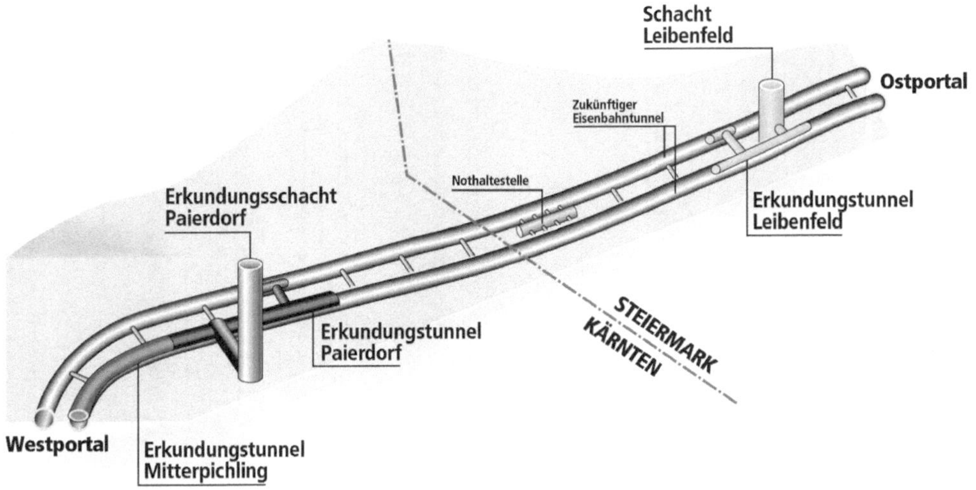

Figure 3. Investigation tunnel concept for the Koralm Tunnel.

step further investigation measures as basis for the detailed design are carried out. For that purpose in the past years a series of deep drillings, reaching depths of 1.160 m, were performed successfully. In the year 2003 the construction of a system of investigation shafts and tunnels with an estimated length of 11 km, as schematically shown in the Figure 3, was started. Depending on the results of the detailed investigation and design as well as the processing by the authorities and the decision how to finance the tunnel project the construction works for the Koralm tunnel could start in the year 2008.

6 SITE INVESTIGATION FOR THE KORALM TUNNEL

The basic target of site investigations shall be a detailed characterisation of the rock mass. Contrary to investigations for many tunnel projects, which are usually performed according to general guidelines, site investigations for the Koralm tunnel were specifically carried out according to the type of rock mass and the phase of the project, Harer & Riedmüller (1999). Specific rock mass related criteria form the basis for rock mass models, generated out of a consistent three-dimensional geological model.

The criteria are defined by lithology and discontinuity structure. They also include the influencing factors in situ and secondary stresses, as well as groundwater. The parameters, which constitute rock mass models, are those, which are relevant for the rock mass behaviour during construction, taking into consideration support requirements and the excavation sequence.

With the objective of cost optimization in mind, the rock mass model, achieved by site investigations is improved stepwise during the different planning phases.

During the phase of route pre-selection (phase 1) the relevant input data based on desk studies and geological mapping of selected areas. At the stage of route selection and environmental impact assessment (phase 2) the geological field work was extended and supplemented by subsurface investigations consisting of core drilling, well log and in situ measurements and geophysical survey. Several improved or new investigation and determination methods Dörrer et al. (2000) were used and led to an excellent, and hopefully realistic, knowledge of the ground conditions (see Figure 2). This formed the basis and supplied the key input data for environmental impact assessment, the assessment for further investigation requirement for the detailed design and tender phase (phase 3) as well as for the risk analysis.

The knowledge of this stepwise investigation procedure is summarized in Geographical Information System (GIS).

7 ROUTE SELECTION

The objective of the first investigation campaign was an engineering geological assessment for a TBM excavation. General investigation targets were the

- bedrock condition (lithologic variations, material properties, permeability, weathering, etc.)
- discontinuities (orientation, spacing, persistency, surface properties and infillings)
- assessment of groundwater conditions
- as well as the identification and characterisation of fault zones by morphological and structural appearances.

Based on the results of the geological mapping the rock mass was classified and a preliminary geotechnical model was developed. Core drillings and geophysical surveys at selected locations with specific geotechnical targets were used to verify that model and to establish a spatial model for each investigation area. Together with the findings of the previous site investigation the "gaps" between the investigation areas were filled and a three-dimensional geotechnical model for the entire route corridor was developed, Goricki et al. (2001), Mussger, et al. (2001) and Steidl, et al. (2001).

Since the portal areas are dominated by environmental considerations and alignment requirements, and only to a limited extent by geotechnical factors, a mostly geotechnically influenced route selection was performed mainly for the crystalline central part of the Koralm tunnel.

The availability of data in the form of a GIS – database allowed a new approach for the ranking of alignment options in the central part.

Out of numerous data sets

- expected radial deformations
- number and length of defined faults
- net drilling rate
- water conditions

were considered to represent the conditions relevant for tunnelling with respect to cost, time and risk.

Spatial information was transformed to a horizontal plane at tunnel level by using the GIS, which represents a generalised model of the ground conditions as a set of map layers and their relationships.

To estimate construction time and costs for different routes the GIS-based system provided a tool, with which the civil engineer was able to select different routes. So the decisions were not mainly based on qualitative assessments but also supported by tangible figures.

The way of defining or refining an alignment by introducing spatial information at the tunnel level in GIS format appeared to be very suitable. It allowed a tunnel alignment to be defined by its major factors – tunnelling conditions expressed in cost and time – and not only by civil engineering or qualitative criteria. However, sufficient and relevant information has to be available at early project stages. Also, sufficient time for planning has to be scheduled to assess all criteria properly.

8 RISK INDUCED ASSESSMENT FOR COST ESTIMATES

In order to achieve economic success for a major infrastructure project like the Koralm Tunnel, difficult decisions based on a wealth of alternatives and unknowns are already required at an early project stage. The GIS-supported geological and geotechnical data also provided important input data for the risk assessments.

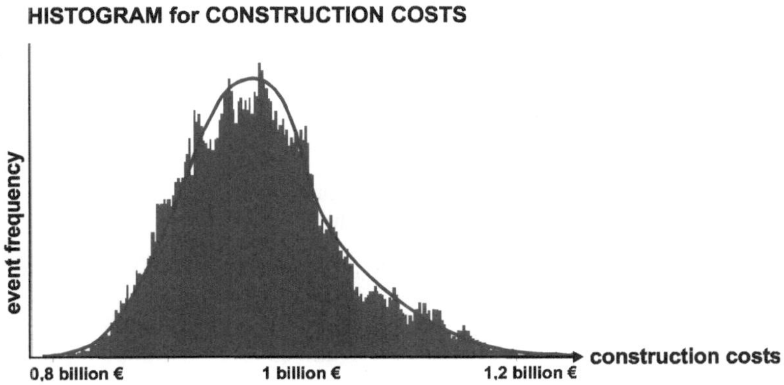

Figure 4. Histogram for construction costs.

One major task in the course of the ongoing design process is the preparation of cost estimates. Considering a great number of risks and uncertainties, variation in cost could be determined in a realistic way to evaluate the risk associated with financing the project; the existence of a consistent geological model provided the basis for geological and geotechnical input needed (see Figure 4).

9 ASSESSMENT FOR FURTHER INVESTIGATION REQUIREMENTS

The existing geological and geotechnical knowledge enabled to work out specific sections with higher geological risk, with special respect to the intended performance of TBM's for the main tunnel tubes excavation. As stated above the crystalline basement is bounded by master, extensive fault zones, which have generated tertiary basins on both sides of the mountain range, as schematically shown in Figure 2. Predominant subsoil and groundwater conditions in these portions (e.g. >120 m water table above tunnel level in loose Tertiary sediments or fault gauges with thickness of several meters in crystalline rocks) might lethally influence TBM-tunnelling if not known in detail and eventually improved in advance of main tunnel excavation.

However, only for specific sections of the tunnel (about 1/3 of the tunnel, see also Figure 3) these additional investigation by means of investigation tunnels is necessary for detailed and tender design. So the application of these time consuming constructions can be reduced to its minimum.

10 CONCLUSION

The portion of investigation with respect of the budget of time and costs must not be neglected. Therefore from the very early stages of planning for the Koralm tunnel a strategy of a stepwise investigation has been executed. The profound knowledge of this stepwise investigation procedure was summarised in an expert system by means of GIS and formed the basis for route selection, environmental impact assessment and risk analysis. For specific sections of the tunnel additional exploration by means of investigation tunnels was found necessary for detailed and tender design.

Performing this way of stepwise improved knowledge enabled to save significantly time and money and though providing sufficient geological data for every project phase.

REFERENCES

Dörrer, Th. et al. 2000. Combined Application of Geophysical Measurements; Case History Koralm Tunnel, *Felsbau* 18 No. 5, pp. 37–47.

Goricki, A., Schubert, W., Steidl, A., Fuchs, R. 2001. Geotechnical assessment of the route corridor, ISRM Reg.Symp. Eurock, Espoo, Finland.

Harer, G., Riedmüller, G. 1999. Assessment of Ground Conditions for the Koralm Tunnel during the Early Stages of Planning, *Felsbau* 17 No. 5, pp. 374–380, http://home.no.net/lotsberg/

Mussger, K., Harer, G., Koinig, J. 2001. Koralm Tunnel – An Innovative Approach to Alignment Selection, *Felsbau* 19 No. 6, pp. 42–47.

Steidl, A., Goricki, A., Schubert, W., Riedmüller, G. 2001. Geological and Geotechnical Ground Characterisation for the Koralm Route Selection, *Felsbau* 19 No. 6, pp. 14–21.

Vavrovsky, G.M, Schneider, K., Harer, G. 2001. Koralmbahn – A new Railway Line in the South of Austria, *Felsbau* 19 No. 6, pp. 8–12.

Geotechnical Risks in Rock Tunnels – Campos e Matos, Ribeiro e Sousa, Kleberger & Lopes Pinto (eds) © 2006 Taylor & Francis Group, London, ISBN 0 415 40005 8

Vibration mitigation at high speed railroads

W. Unterberger
iC consulenten ZT GmbH, Vienna, Austria

ABSTRACT: Noise and vibration are among the most important environmental issues to be addressed when designing high-speed railroad lines. While noise prognosis follows well established design procedures, vibration prognosis and mitigation is a much younger field which is still very much open to innovation. In Austria, a system of vibration prognosis has now been practiced and successfully applied for several years at a number of high-speed railway projects that relies on a combination of measurements, analytical approaches and numerical analysis. The prognosis system is currently being expanded to include probabilistic approaches.

1 CAUSES AND EFFECTS OF RAILROAD VIBRATIONS

Vibration engineering has become an indispensable part of the design of any new or upgraded railway. Increased environmental awareness and stricter health regulations have furthered this development. While noise prognosis and mitigation is a well developed and highly standardised field, vibration prognosis and mitigation is still developing new approaches and solutions. Since vibrations propagate through the ground, all the unknowns and unknowable of geotechnical engineering apply to this field.

Train vibrations are caused at the contact wheel – rail. They are significantly affected by factors such as train speed, axle weight, quality of rolling stock, quality of rail/sleeper superstructure and stiffness and quality of subgrade. Attempts have been made to deterministically model the origin of rail vibrations and significant parts of a vibration signal can be attributed to definite causes in the train – track-bed system such as sleeper spacing, natural frequencies of locomotives and coaches etc. Still, each train has a typical emission signal on each section of an alignment. This means, that measurements remain the best way for determining vibration emissions.

The amplitude and frequency content of the vibration signal are modified as the wave propagates outwards from the source. Railroad structures such as cuts, embankments, retaining walls and especially tunnels have a strong influence. Ground layers may reflect or refract the waves, sometimes blocking them, sometimes providing preferential transport media. The groundwater table plays a strong role. The vibrations then impact the foundations of the buildings and are further modified depending on the total mass of the building and the natural frequencies of the ceilings. Overall, vibration mitigation is an attempt to assess the influences of these various factors, especially regarding their influence on signal frequencies.

2 ASSESSMENT OF EFFECT OF VIBRATIONS ON HUMANS

Vibrations affect humans in two ways: as recognizable vibrations or as "hearable" secondary noise, which is re-radiated by the walls and ceilings of a room. Vibration velocity alone is not sufficiently accurate as a means of assessing the effect of vibrations on humans. The so-called K-value is an internationally recognised factor – basically a frequency weighted effective value of the time – velocity signal. While frequency weighting is standardised for the K-value internationally in ISO standards, the method of calculating the effective value varies from standard to

Table 1. *K*-values.

0.1	Not perceptible
0.2	Just perceptible
0.4	Slightly perceptible
0.8	Perceptible
1.6	Clearly perceptible
6.3	Strongly perceptible
100	Very strongly perceptible

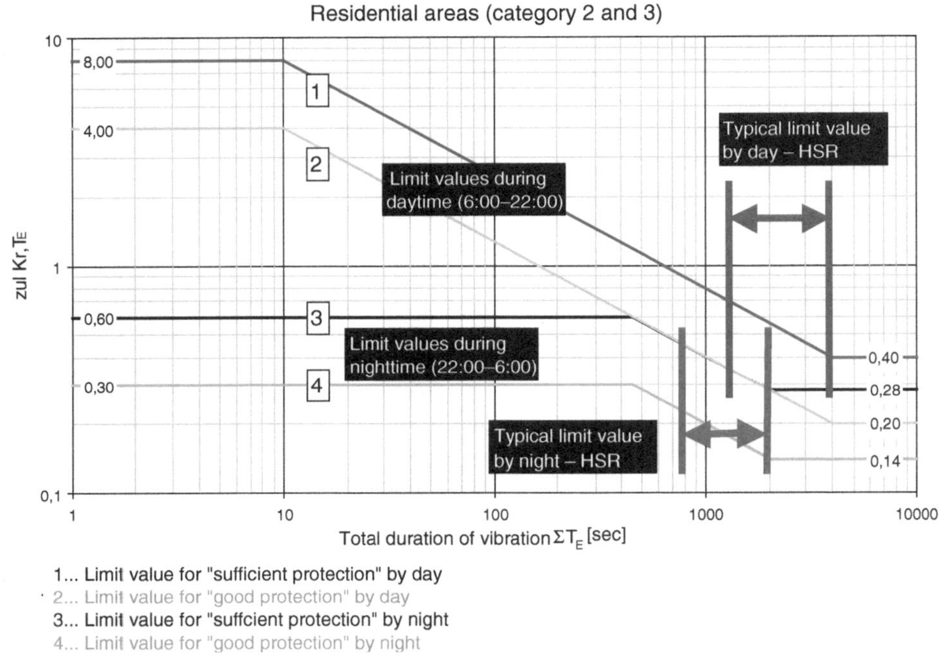

Figure 1. Limit values.

standard (and from application to application). The factors that affect effective value calculation are the length of the integration time window and the possible use of a "memory function," which weighs more recent values less than more recent values.

K-values can be related to subjective perception as outlined in Table 1.

Limit values generally aim at the range between $K = 0.2$ and 0.4. Austrian standard Önorm S 9012 determines the limit values for train vibration according to the total duration of train passages.

Limit values also differ according to the zoning of the area (residential, commercial, industrial) and according to day and night time. Two levels of protection are defined: "good" and "sufficient" protection. "Good" protection is usually aimed for when constructing new lines, "sufficient" protection when rebuilding and expanding existing lines.

Secondary noise is measured in dB [A]. However, it is usually not measured directly as noise but is determined indirectly from the vibrations of ceilings and walls, since it is very hard to separate secondary noise from other noise sources.

3 POSSIBLE COUNTERMEASURES

Countermeasures fall into three major categories.

1. Measures close to the source
2. Measures in the propagation path
3. Measures at the buildings

Measures taken close to the source are usually most efficient. However, since they affect the trackbed system, they have to be carefully chosen to conform to other requirements of the rail-sleeper-ballast system such as stiffness, train dynamics, maintenance etc. Measures at the source usually are a combination of an elastic element and a massive element. Examples are:

- Sub-ballast mats, where the mass is the ballast or slab track system
- Mass spring systems
- Padded sleepers, where the mass is constituted by the sleepers and rail

Each of those systems has a typical natural frequency. Vibrations are increased in magnitude at frequencies around the natural frequency and are reduced at frequencies starting at about 1.4 times the natural frequency. The system therefore has to be carefully chosen to correspond with the critical frequencies of the surrounding buildings.

Measures in the propagation path include all measures taken at tunnels by strengthening the linings. It is well known that a more massive invert has a very significant effect on vibration levels. At surface lines, dewatering trenches and retaining walls have a mitigating effect. Special solutions include purpose-built gas-cushion slots – a method developed in Sweden.

Measures at the buildings have to be taken with care, since a modification of the bearing structure of a building can shift the natural frequencies and actually cause an increase of vibrations. However, sophisticated systems exist that permit elastic bedding of complete buildings.

4 PROGNOSIS AND DESIGN

Vibration prognosis and mitigation design has to address the interdependent and sensitive aspects outlined above. The methodology rests on four classical pillars:

1. Measurements
2. Analytical approaches
3. Numerical approaches
4. Experience and rule of thumb

Of the four pillars, numerical approaches have been applied only recently, since dynamic numerical analysis is significantly more computer intensive than conventional numerical analysis. No method can succeed on its own, a combined design approach is needed. Measurements are best suited for determining emission spectra, natural frequencies and other dynamic properties of buildings and for verification of prognoses. Analytical approaches can be used for determining building properties when the number of buildings is too large to measure each building and are ideally suited for estimating the effect of countermeasures such as mass-spring systems. Numerical methods come into their own when large modifications are made to the existing situation, especially in case of tunnel construction or large embankments and dams. They are also well suited to the modelling of complex, new buildings constructed close to rail lines – a case that is very common in metro construction. Experience – as usual – is needed at all stages of the design process.

5 NUMERICAL METHODS – VARIATION OF GROUNDWATER LEVEL

Since numerical methods are a new element of vibration design, a closer look at their merits is justified. The following diagrams are the results of two series of analyses that were performed specifically to show the power of parameter studies using numerical tools. They were performed using the Finite Difference Code FLAC 4.0.

Figure 2 shows the model used, a tunnel in granular soft ground with the groundwater layer situated at various positions. The following plots use a specific method of assessing the results of dynamic analyses at a glance. Kinetic energy density is tracked over the duration of the simulation and summed up for each node. This total kinetic energy density is then plotted, thus showing areas of strong vibrations and areas of reduced vibrations. The results of various parameter studies can thus be easily compared.

Figure 3 shows the first variation, the groundwater level is on tunnel diameter below the invert. Pink, yellow and green shows high kinetic energy, blue hues denote low kinetic energy. A change in color corresponds to a decrease in vibration energy by a factor of 2. The groundwater level acts as a reflector, energy is high around the tunnel and at the surface and low in the groundwater.

Figure 4 shows the situation with a higher groundwater level. The vibration situation deteriorates slightly, since the groundwater level is closer to the tunnel and the surface, vibrations at the surface are larger.

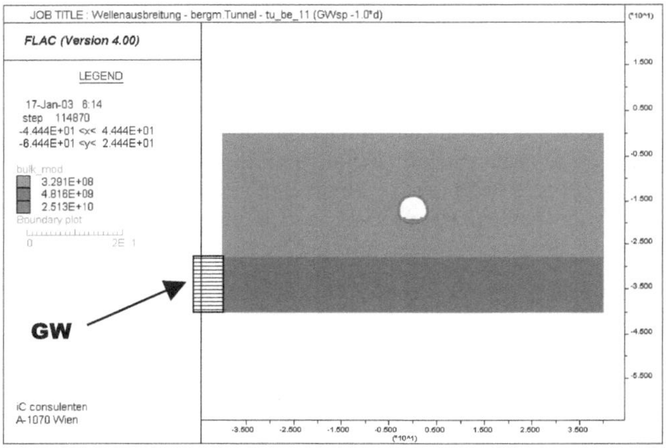

Figure 2. Calculation model, tunnel in granular ground with ground water level below invert.

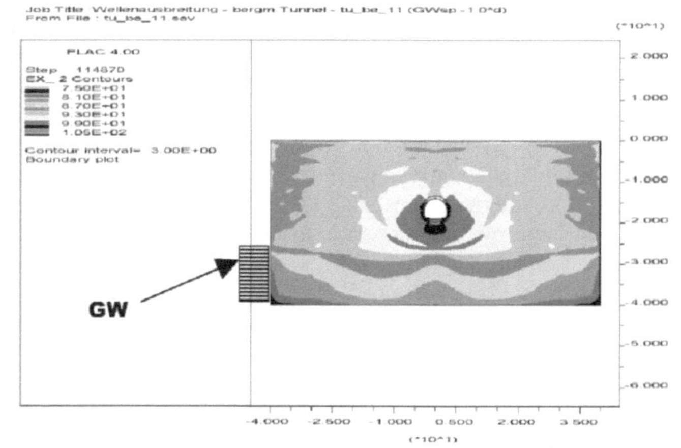

Figure 3. Kinetic energy density, groundwater one tunnel diameter below invert.

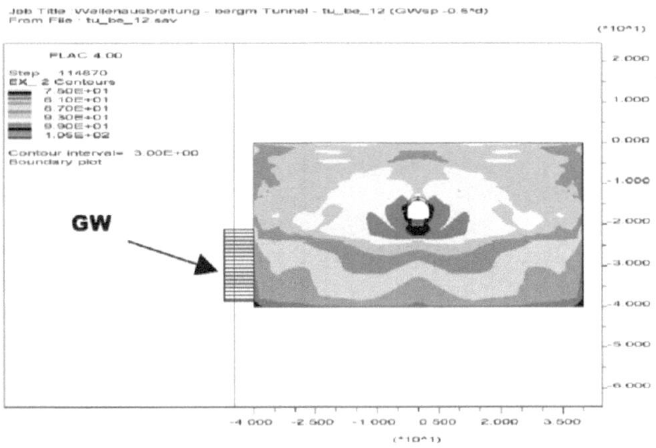

Figure 4. Kinetic energy density, groundwater half a tunnel diameter below invert.

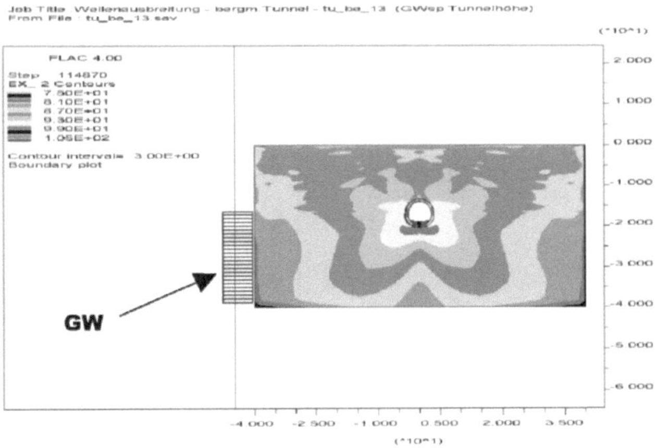

Figure 5. Kinetic energy density, groundwater at tunnel crown.

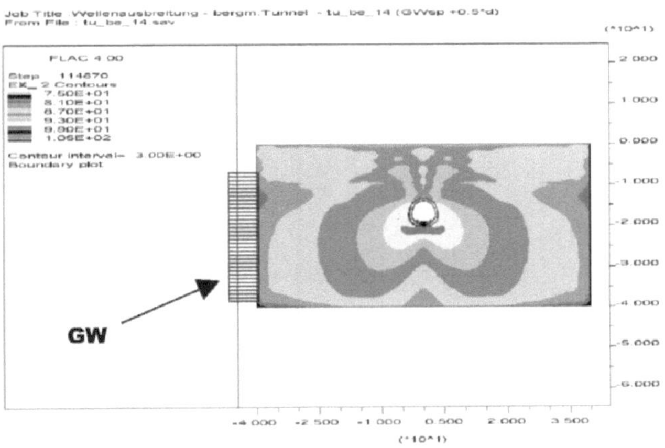

Figure 6. Kinetic energy density, groundwater above tunnel crown.

When the tunnel is immersed in the groundwater, as shown in Figure 5, a dramatic change in the vibration situation is caused. The groundwater no longer acts as a reflector but actually provides a preferred propagation path for the waves downwards and away from the surface. Surface vibrations are thus strongly reduced.

If the groundwater level is raised further as shown in Figure 6, vibration levels at the surface are reduced yet again.

6 NUMERICAL METHODS – VARIATION OF RAILWAY ALIGNMENT

Similar studies were performed on the effects of a railway line gradually being lowered from an embankment situation until a tunnel situation is reached. Kinetic energy density plots are used again. Figure 7 shows the situation where the line is situated on a dam, Figure 8 shows the case

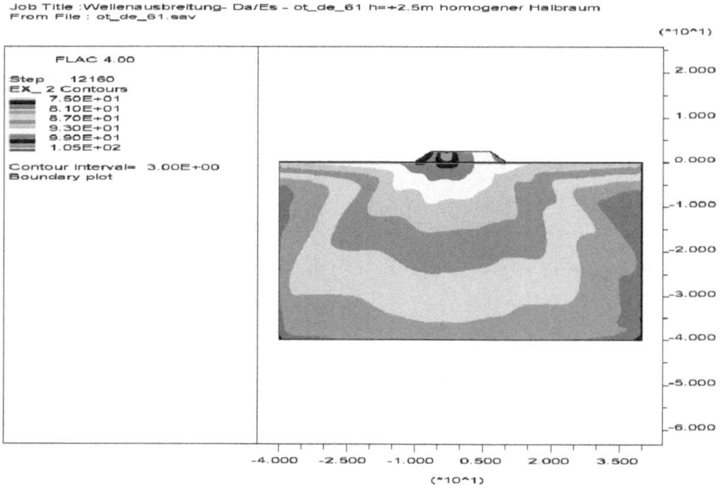

Figure 7. Kinetic energy density, track on dam.

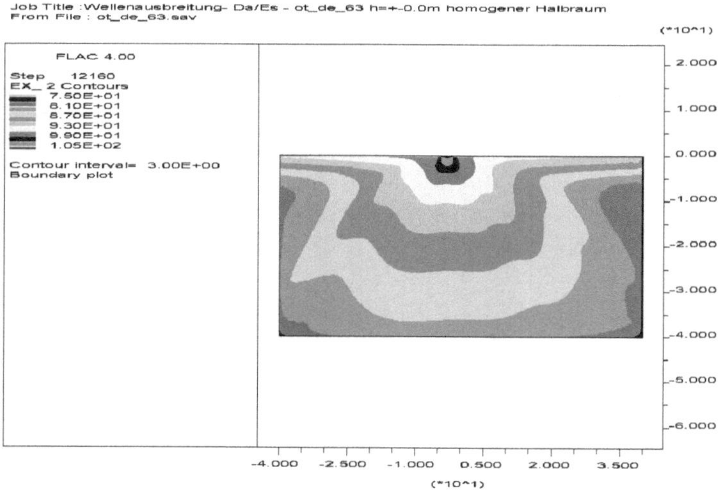

Figure 8. Kinetic energy density, track at grade.

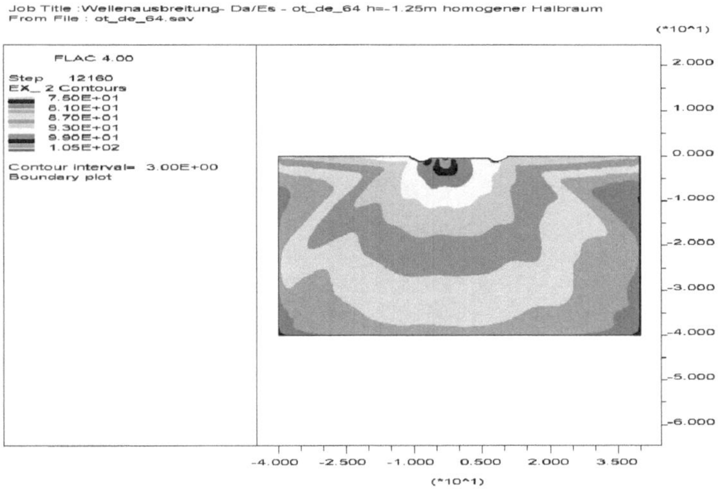

Figure 9. Kinetic energy density, track at grade with drainage ditches.

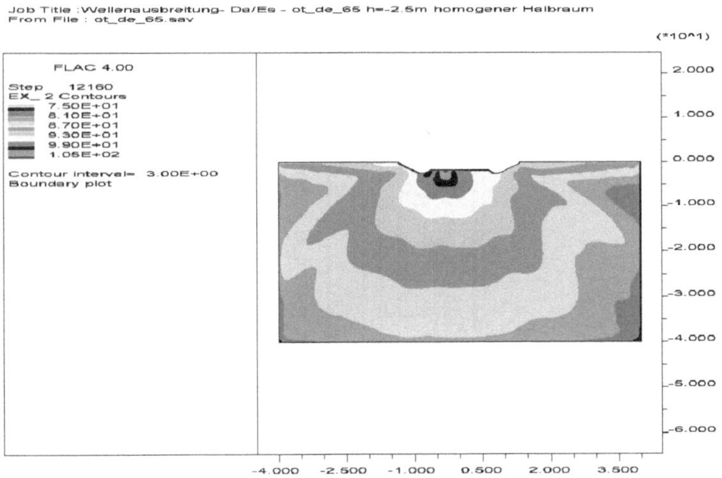

Figure 10. Kinetic energy density, track in cut.

where the line is at grade. The yellow zone can be seen to extend further out in Figure 8, indicating higher vibrations. The mass of the dam serves to reduce vibrations in the case shown in Figure 7. Figure 9 shows a single, very interesting modification. Drainage ditches are modeled and vibrations are reduced. This is due to the fact that the ditches interrupt the formation of the surface "Rayleigh" waves and thus reduce vibrations.

Figures 10, 11 and 12 show the effect of gradually lowering the alignment further, Figure 10 represents a cut, Figure 11 a ramp to a tunnel section and Figure 12 a cut-and-cover tunnel. Vibrations decrease, with a clear jump due to the concrete structure of the ramp and an even stronger jump when the tunnel provides a very stiff structure.

In addition to the technical information provided by these studies, it also becomes clear that appropriate visualisation techniques are essential for successful evaluation and interpretation of calculation results.

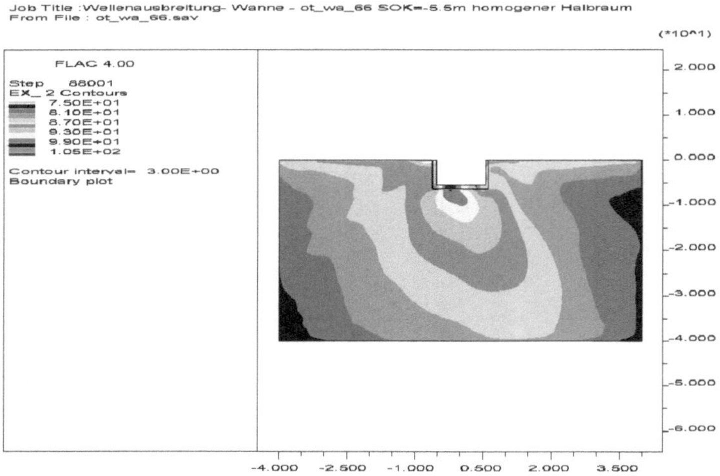

Figure 11. Kinetic energy density, track in ramp to tunnel.

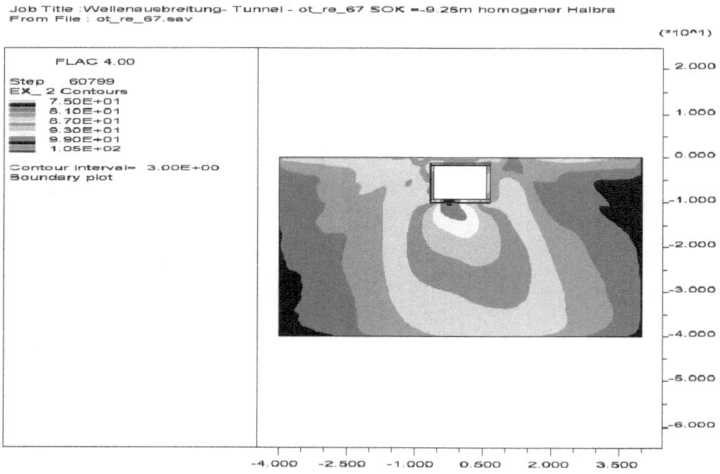

Figure 12. Kinetic energy density, track in tunnel.

7 SUMMARY AND CONCLUSION

Vibration prognosis and design of mitigation measures involve a design process that is based on measurements, analytical approaches, numerical methods and experience. Numerical methods have seen a strong growth in the field in recent years and now form an important part of a modern design approach. Successful vibration design depends on the careful combination of all methods mentioned above.

Geotechnical Risks in Rock Tunnels – Campos e Matos, Ribeiro e Sousa, Kleberger & Lopes Pinto (eds) © 2006 Taylor & Francis Group, London, ISBN 0 415 40005 8

Evaluation of the reliability in reference geological hydrogeological models

Guido Venturini, Paolo Perello, Antonio Dematteis, Gianpino W. Bianchi, Luca Delle Piane & Antonio Damiano
SEA Consulting S.r.l., Turin, Italy

ABSTRACT: Evaluation of the reliability in reference hydrogeological models (RHM) represents an essential part to be defined during design and/or construction of deep tunnels, in order to correctly evaluate technical and financial risks as well as environmental hazards due to the potential interaction between underground works and surface water resources (springs, lakes, etc). Deep tunnels in alpine-like mountain belts, generally crosscut a low primary permeability bedrock so that the main groundwater flow is normally related to brittle tectonic structures like faults, cataclastic shear zones, thrusts and joint systems. Therefore, the RHM reliability directly depends on the Reference Geological Model (RGM) and particularly on the quality and reliability of the geological data which are at the basis of the RGM.

The reliability of such data depends on A) the detail and reliability of geological, structural and hydrogeological surveys carried out on the surface, B) the outcrop percentage, the availability of key-outcrops and the structural complexity, C) the persistence and the degree of evolution of the existing brittle structures (faults, thrusts, joint sets, etc.), D) the amount and type of available investigations, E) the depth of the underground excavations, F) the possibility of projecting geological borehole data to the working level and the difficulties in the interpretation, G) the type and amount of borehole hydraulic tests and H) the possibility of determining the recharge area and the surface hydraulic balance.

1 INTRODUCTION

In Italy and the neighbouring countries, several deep high-speed/high-capacity railway tunnels are at present being constructed, crossing the Alpine chain and the Apennines, and are the subject of detailed geological studies. The SEA Consulting staff is involved in some of these projects, such as the 53 Km Maurienne–Ambin and the 8 Km Bussoleno tunnels of the new Lyon–Turin link, the 38 Km "Galleria di Valico" of the Milano–Genova high speed link and the Italian side of the 54 Km Brenner Base Tunnel (BBT), as contractors for geological and hydrogeological studies and with tasks of construction supervision.

In facing the problems related to the planning of these tunnels, SEA Consulting could rely on the experience acquired during the planning and/or construction of several water diversion tunnels and underground power plants, located in the western and central Alpine belt, where very difficult conditions related to underground water were expected and/or encountered (Figure 1).

The hydrogeological risks related to unexpected heavy water inflow and drainage of surface resources represents, together with the occurrence of unexpected geological and geomechanical conditions, one of the most challenging and troublesome aspects related to tunnelling. In some cases, strong water inflows, locally exceeding 50 l/sec (Figure 2), caused a break in the excavation; the delay caused in turn heavy losses of money and in some cases gave way to onerous claims due to the underestimation of such problems in tender documents.

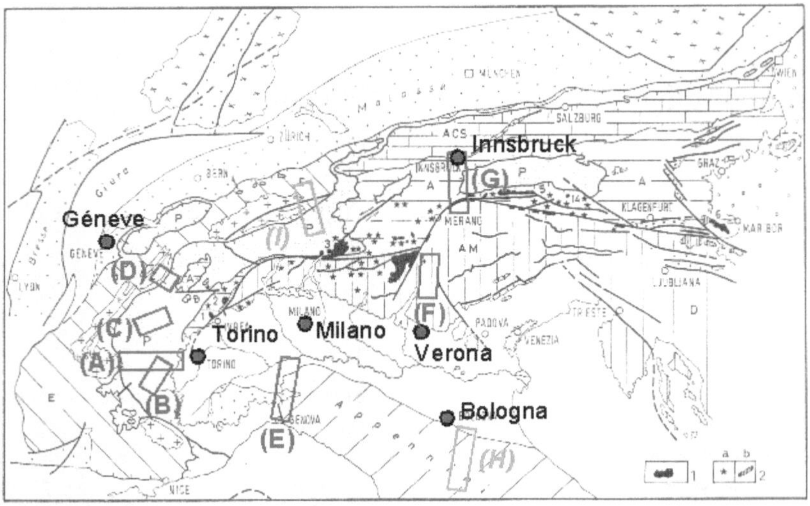

(A)	**Lyon/Turin railway tunnel**	**65 km (2,5 km)**	**8.200**	**M Euro**
(B)	**Pont Ventoux HP plant**	**20 km (0,8 km)**	**200**	**M Euro**
(C)	Orco Valley HP plants	25 km (0,7 km)	350	M Euro
(D)	Mont Blanc Highway tunnel	12 km (2,6 km)	320	M Euro
(E)	**Milan - Genoa railway tunnel**	**38 km (0,7 km)**	**1750**	**M Euro**
(F)	**Brenner south railway link**	**32 km (0,7 km)**	**1250**	**M Euro**
(G)	Brenner Basis Tunnel	52 km (1,3 km)	3800	M Euro
(H)	*Bologna - Florence railway tunnels*	*40 km (0,6 km)*	*4500*	*M Euro*
(I)	*Gotthard Basis Tunnel*	*54 km (1.3 km)*	*4200*	*M Euro*

Figure 1. Location of the main underground works where SEA Consulting is presently involved, locally in J/V with other consulting companies (bold). Length, overburden and estimated cost for each tunnel are listed below the figure.

Figure 2. Lyon–Turin railway tunnel, western Alps: water evacuation problems in the downstream driven Modane adit to the base tunnel (A); serious water inflow from a probe drilling (B).

At the same time, systematic drainage of the fractured and faulted rock mass and consequently of some important surface water resources caused in many cases serious environmental damages, with strong and persistent interference with the reservoirs being exploited by public aqueducts (Figure 3); this led in some cases, to legal proceedings as, for instance, for the Fiorenzuola Tunnel of the Bologna–Florence high-speed line.

For these reasons, in the last five years hydrogeological studies concerning the main tunnels under construction in Northern Italy were implemented, in order to quantify as precisely as possible the order of magnitude of the water amount to be drained by the tunnel, and to design measures able to reduce as much as possible the interference with the surrounding water reservoirs.

These studies are based, above all, on the execution of a Reference Geological Model (RGM) which is the best one consistent with the amount of knowledge available for a given project (Delle Piane et al., 2001; Venturini et al., 2001; Perello et al., 2003); the Reference Hydrogeological Model (RHM), is then focused on the RGM and is the result of hydrogeological observations, analyses, monitoring and tests applied to the RGM.

It is of the outmost importance to evaluate the effective reliability of the proposed RGM/RHM, in order to provide clients, engineers and contractors with a tool to quantify the technical and financial risks related to this subject.

In this contribution we propose a methodology to quantify the reliability of the RGM/RHM by means of a number of parameters that can easily be determined by analysing the methodological approach that leads to define the hydrogeological uncertainties within a given area. This method was developed during the studies for the 10 Km Venaus investigation tunnel of the Lyon–Turin link (Italian Western Alps, Figure 4) and the southern link to the BBT (Italian Eastern Alps, Figure 6), and during the construcion supervision of the Modane Adit to the underground Modane station of the Lyon–Turin line (French Western Alps, Figure 5).

The method is still in course of development and will certainly require further implementations, derived from new projects and from the construction back-analysis of tunnels previously investigated as well.

Figure 3. Pont Ventoux hydro-power plant, Susa Valley, Western Alps – Italy: water drainage reaching 130 l/sec, causing strong an permanent interference with drinking water springs in the surrounding villages.

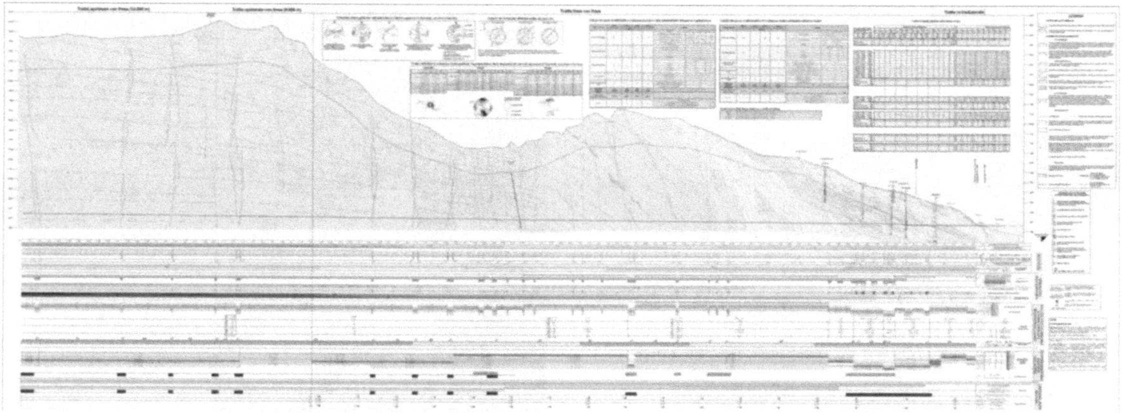

Figure 4. Geological cross section of the Venaus investigation tunnel. Several water inflows under 2000 meters overburden are expected in the final part of the tunnel which is especially planned to understand in detail the underground water conditions as well as the tensional state of the rock mass.

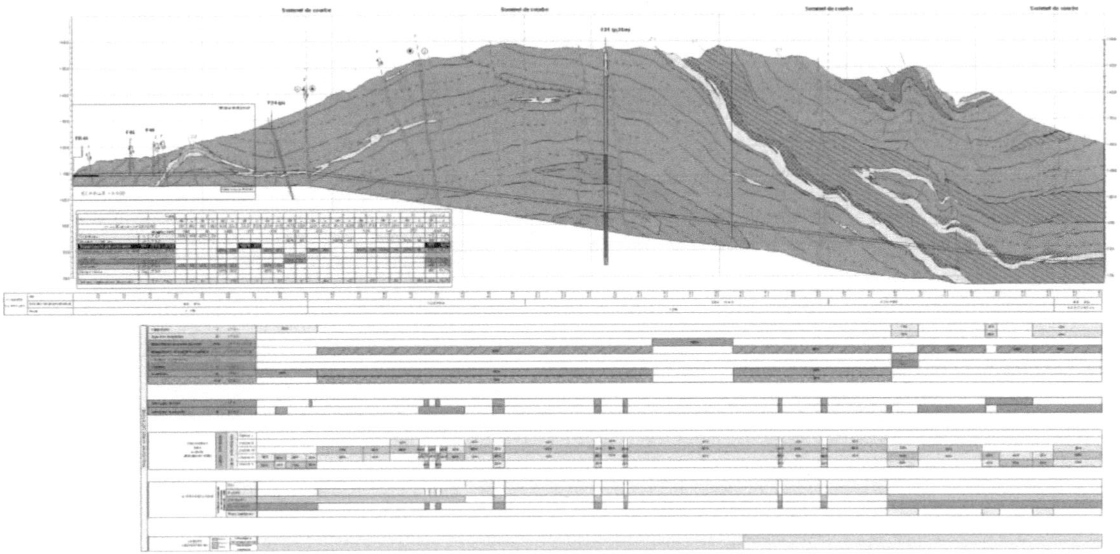

Figure 5. Geological cross section of the Modane adit. At present, water inflow in this tunnel reaches 180 l/sec at 7 bars pressure.

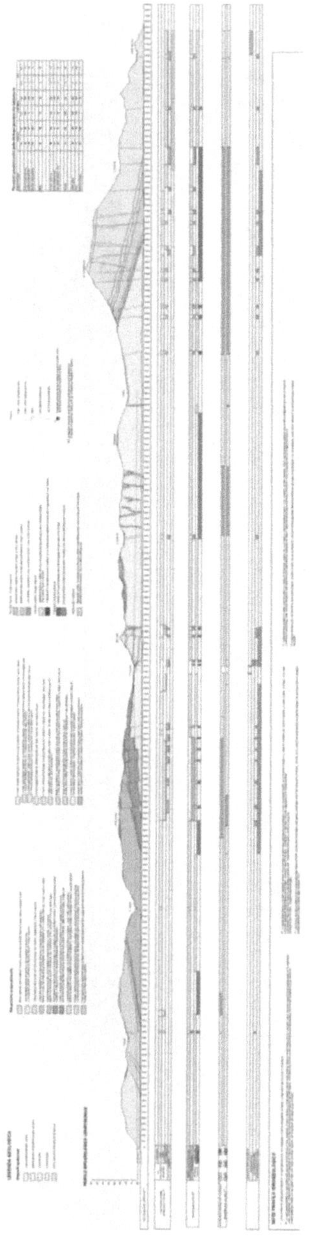

Figure 6. Hydrogeological cross section of the southern link to the Brenner Base tunnel (Italian Eastern Alps).

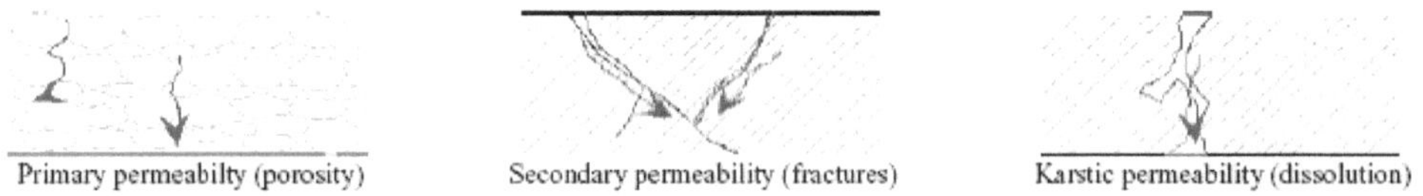

Figure 7. Sketch of permeability in crystalline rock terrains.

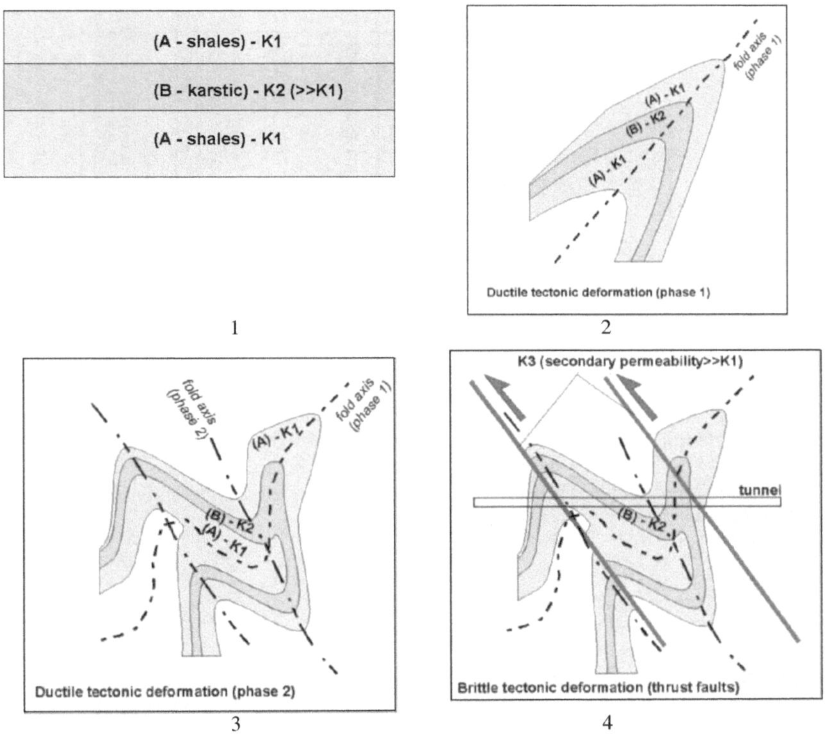

Figure 8. Example of reconstruction of a polyphasic deformation history affecting rock units with different values of k. A couple of hydrogeological units (A and B, 1) is firstly deformed by a tight fold (2); a second folding phase (3) can change the phase 1 geometry, by redistributing into space the permeable and impermeable units, complicating the previous setting. In the development of a mountain belt, the end of the ductile deformation is generally followed by faulting and/or thrusting (3) that generate new permeability channels. Note that the occurrence of karstic permeability is a secondary feature, which develops after ductile deformations.

2 THE HYDROGEOLOGICAL MODEL AS DEPENDING FROM THE GEOLOGICAL SETTING

Water circulation depends on primary permeability, related to the original rock porosity and by a secondary permeability generated by brittle deformation (Figure 7). Primary permeability can be generally considered very low in crystalline rocks, while secondary permeability related to fracturing may be responsible for severe water inflow.

In order to reconstruct the 3D distribution of layers or rock units having different permeability coefficients (k), it is extremely important to define first of all a reliable structural model, by

A

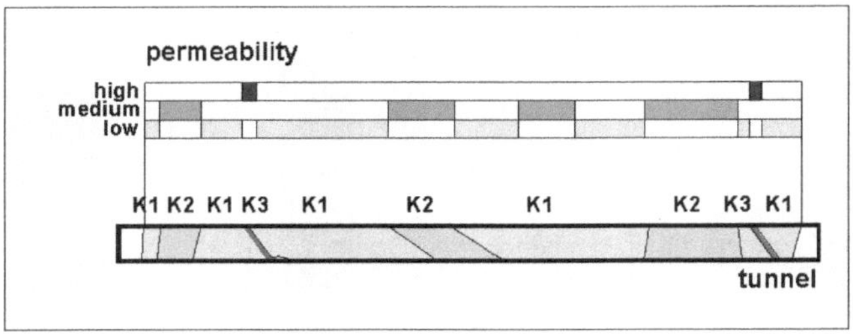

B

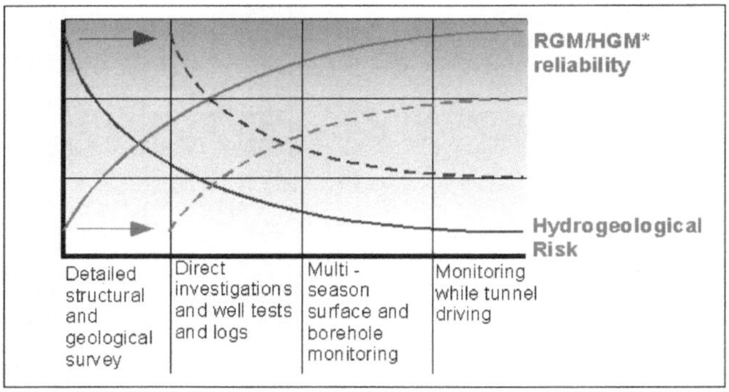

Figure 9. Hydrogeological reconstruction along a tunnel cannot be carried out without a preliminary definition of the whole lithostratigraphic and tectonic (brittle and ductile) contest (A). Hydrogeological risk can be minimised only by increasing the RGM/RHM reliability. Note that when a reliable geo – structural model is lacking (e.g. no detailed geostructural data available), it will be impossible to reach a low hydrogeological risk level.

reconstructing the different deformation steps (e.g. folding sequence) and the geometry of the lithostructural units which influences, in turn, the spatial distribution of the hydrogeologic units (Figures 8 and 9).

3 RGM–RHM RELIABILITY

Having defined the importance of the RGM, several elements have to be taken into account to evaluate its reliability and, therefore, the reliability of the derived RHM. They are:

1. type of surface geological surveys;
2. scale(s) of the geo-structural survey;
3. percentage of outcropping bedrock and presence of key-outcrops for the understanding of structural relationships;
4. ductile structural complexity (e.g.: multiphase folding);
5. brittle structural complexity;
6. location of the boreholes, when available, with respect to the underground works;
7. type and quality of the available boreholes;
8. type and quality of borehole tests;
9. tunnel overburden.

	1	2			3	4	5		6	7	8	9	
	Geo-structural survey scale	**type of survey realised from the surface**			**Percentage of outcropping bedrock**	**Ductile Structural complexity**	**Brittle structural complexity**		**Location of the boreholes with respect to the alignment**	**type and quality of the available boreholes**	**type of borehole tests**	**tunnel overburden**	
		Geological survey	Structural survey	Hydrological survey			Persistence (*surface trace length vs. tunnel overburden average*)	Development	far from the alignment *difficult projection)*			difficult projection	easy projection
Reliability Increasing	1/25.000	Litho-stratigraphic	From aereo photos	spot sampling	0 - 40%	three transposed deformation phases	<1	scattered systems	far from the alignment *easy projection)*	100% destructive (cuttings) without logs	No test	>1000	
	1/10.000	geo – structural (*semplified*)	Geometric	Systematic monitoring	40 - 70%	three deformation phases	=1	moderately pervasive and developed systems	close to the alignment *difficult projection)*	100% destructive (cuttings) with logs	geophysics logs only	300- 1000 m	>1000 m
	1/5.000	geo – strctural (*Detailed*)	Geometric and kinematics	Multi - season monitoring	70 - 100	two deformation phases	>1	pervasive and developed systems	close to the alignment *easy projection)*	50% destructive (cuttings) with logs	geophysics logs + hydraulic tests	100 - 300 m	300 - 1000 m
						One deformation phase			on the alignment	destructive with BHTV and structural study of the borehole wall	geophysics logs + hydraulic tests + water sampling	0 -100 m	100 - 300 m
						No deformation				Cored boreholes	geophysics logs + hydraulic tests + water sampling + microflow		0 - 100 m

Figure 10. Detailed schedule of the parameters that have to be determined in order to define the reliability of the RGM/RHM. The method permits to quantify the increasing or decreasing influence of each parameter on the reliability. At present, the most difficult and subjective step is to determine the interaction between the different parameters (e.g. influence between borehole location and tunnel overburden).

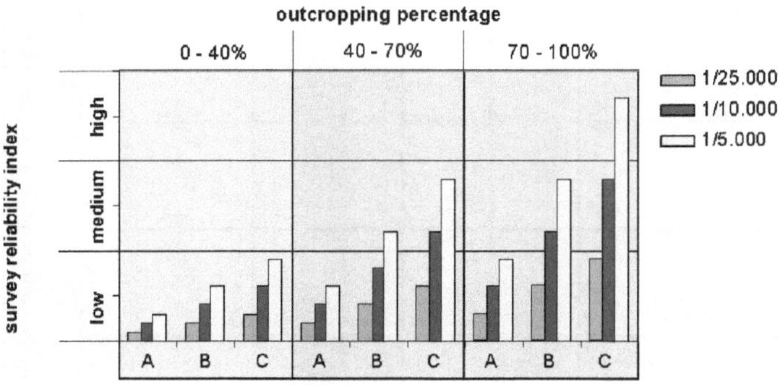

A) Lithostratigraphic with structural analysis from aeral photos
B) Lithostratigraphic and geostructural with geometric analysis
C) Lithostratigraphic and geostructural with geometric and kinematic analysis

3.1 *Survey reliability index*

One of the most important items that play a role in determining the reliability of the RGM/RHM is the type of geological survey carried out during the studies. The graph below correlates the three main elements related to the geological surveys:

1) scale of survey (from 1/25.000 to 1/5.000);
2) outcrop percentage;
3) type of survey (lithostratigraphic without structural analysis, lithostratigraphic and structural with geometric analysis, or detailed structural survey).

As can be seen in the graph below, a small scale survey (e.g. 1/25.000) usually allows to reach only a low reliability index, due to the fact that at this scale no detailed structural analysis can be done and the reconstruction of the geometry of geological units and of the relationships between ductile and brittle deformation are necessarily limited to a general overview.

4 CONCLUSIONS

The method to determine the reliability of the RGM/RHM is developed starting from a considerable numbers of tunnelling case histories from the Alps. The method, to be improved by analysing the relationships between forecasts and observed data, has the advantage that information can be easily collected and analysed and can be experimentally applied to different case studies, improving at the same time the feedback.

According to the case histories where such method was applied, the reliability of the RGM/RHM for a given geological setting can be improved in different ways, e.g. by increasing the number of boreholes and of borehole tests. Nevertheless, a proper preliminary geo-structural survey may lead to a greater increase of the model reliability, since it gives basic information to develop a model and it allows to better interpret the other data sets (e.g. borehole data). Moreover, geo-structural surveys represent the only means to describe complex tectonic settings and the related geometries.

From the suggested analysis it appears that, even if the outcrop conditions are very poor (i.e. <40%), a detailed surface survey permits to increase the RGM/RHM reliability and to decrease the geological/hydrogeological risk.

Geotechnical Risks in Rock Tunnels – Campos e Matos, Ribeiro e Sousa, Kleberger & Lopes Pinto (eds) © 2006 Taylor & Francis Group, London, ISBN 0 415 40005 8

Author Index